Mission to Neptune

Case Study of an Unmanned Robotic Mission to Neptune

By

Dr Ugur GUVEN

Guven Publications
Miami-Paris-Istanbul

Disclaimer

Although the author and publisher have made every effort to ensure that the information in this book was correct at press time, the author and publisher do not assume and hereby disclaim any liability to any party for any loss, damage, or disruption caused by errors or omissions, whether such errors or omissions result from negligence, accident, or any other cause.

Copyright

Mission to Neptune: Case Study of an Unmanned Robotic Mission to Neptune

© April 2023, by Dr. Ugur GUVEN
Published through Amazon under Guven Publications

ASIN:

ISBN: 9798390098752

Miami-Paris-Istanbul
Contact: drguven@live.com

Foreword

As a Rocket Scientist, I have worked on many research papers and textbooks, and they are sold across the world. I have especially worked on interstellar travel techniques and if you go to my website, you can see and read many of the research that I have worked on. However, I believe that the most important challenge ahead of us is exploring our solar system. I already have written books for several destinations in the solar system including Moon, Mars, Jupiter, and Europa.

This book is the continuation of my Solar System series as it focuses on a mission to Neptune. In a way, Neptune is an interesting destination and unlocking the secrets of Neptune can pave the way for a greater understanding of our solar system and Earth. Neptune is an ice giant and its atmosphere made of hydrogen, helium with methane gives a special blue color to it. It's one of the most windiest planets in the solar system with winds having speeds over 2000 km/h. and it's the only planet that can't be seen with the naked eye. It is the coldest planet in the solar system and interestingly like Saturn it contains rings though they are not as prominent.

I dedicate this book to pioneers of space Sergei Korolev and Werner von Braun who made space travel possible.

Prof. Dr. Ugur GUVEN
www.drguven.com

SUMMARY

The aim of this Bok is to study different categories of robotic space probes and present a case study of an unmanned robotic mission to planet Neptune. A detailed analysis of the mission requirements will provide a general idea about the types of subsystems and on-board scientific instruments that this spacecraft would carry in order to explore Neptune. The proposed mission is an orbiter or a flyby which would examine the planet Neptune from a distance. Hence a detailed analysis on the trajectory followed by the spacecraft from Earth to Neptune is presented here.

Contents

LIST OF TABLES 3

LIST OF FIGURES 4

1.INTRODUCTION 8

   1.1.   The need for robotics in space...9

2.SPACE ROBOT SPECIFICS 11

   2.1. Machine Intelligence 11

   2.2 General Classification of Scientific Spacecraft...12

   2.3. Functional Subsystems 17

      2.3.1. Structural subsystem...17

      2.3.2. Power subsystem 18

      2.3.3. Attitude control subsystem 19

      2.3.4. Thermal control subsystem..20

   2.4. Telecommunications 21

3.Flyby Spacecrafts..22

   3.1. Gravity assist (or the mechanics of a flyby)..23

   3.2. Phases of a flyby mission 25

      3.2.1. Observatory Phase 25

      3.2.2. Far Encounter Phase 25

      3.3.3. Near Encounter Phase:...26

      3.3.4. Post Encounter Phase..26

   3.4. Important Flyby Missions 27

      3.4.1. Mariner 10 27

      3.4.2. Pioneer 10 and 11 29

3.4.3. Voyager 2 32

4.Neptune and General Spacecraft Configuration 34

4.1. Neptune 34

4.1.1. Surface and Weather Conditions on Neptune: 34

4.1.2. Exploration of Neptune: 35

4.2. Spacecraft Instruments 38

4.3. Propulsion Subsystems...40

4.3.1. Ion Propulsion..42

4.3.2. Ion Thruster Operation 43

4.3.3. The Electric Propulsion System 44

5.STATE VECTOR 49

5.1. The Two Body Problem 49

5.1.1. Equations of motion in Inertial frame of reference 49

5.1.2. Equations of Relative motion 50

5.2. Types of Orbits...52

5.2.1. Elliptical Trajectories (0<e<1) 52

5.2.2. Parabolic Trajectories (e=1) 54

5.2.3. Hyperbolic Trajectories (e>1) 55

5.3. Orbits in Three Dimensions 58

5.3.1. Orbital Elements 59

5.3.2 The Perifocal Frame 62

5.3.3. Transformation between Ecliptic and Perifocal frames 63

5.4. Planetary Ephimeris 68

7

6.Interplanetary Spacecraft Trajectories 70

   6.1. Sphere of Influence and the Method of Patched conics 71

   6.2. Planetary Departure...74

   6.3. Planetary Arrival 77

   6.4. Orbit Determination: Lambert's Problem 79

CONCLUSION...84

APPENDIX:..87

LIST OF TABLES

Page

Table 1: General data for planet Neptune 35
Table 2 : List of Scientific Instruments on-board Voyager 2 39
Table 3: Equation for elliptical orbits 53
Table 4: Hyperbolic trajectory equations..56
Table 5: J2000 orbital elements and centennial rates 59

LIST OF FIGURES

Page

Figure 1.1: Different planetary flyby's 14
Figure 3.1: Gravity Assist 23
Figure 3.2: Mariner 10 27
Figure 3.3: Power subsystem – Mariner 10 27
Figure 3.4: Pioneer 11 30
Figure 3.5: Scientific instruments aboard Voyager 2 31
Figure 4.1: Atmosphere of Neptune...34
Figure 4.2: Various spacecraft instruments 38
Figure 5.1: Force of Gravitation 48
Figure 5.2: Elliptical Orbit..51
Figure 5.3: Eccentricity anomaly...52
Figure 5.4: Parabolic Orbit 54
Figure 5.5: Hyperbolic orbit 55
Figure 5.6: Planetary ephemeris 57
Figure 5.7: The perifocal frame 62
Figure 5.8: Transformation of frames of reference...63
Figure 5.9: Rotation of axes...64
Figure 6.1: Decreasing gravitation with increasing radius 71
Figure 6.2: Planetary Departure..74
Figure 6.3: Planetary arrival 77
Figure 6.4: Interplanetary trajectory 79

Chapter 1: INTRODUCTION

A robot is any computer controlled electro-mechanical machine, which performs mechanical tasks at human command. These tasks can be repetitive in manner (for example, industrial robots) or may require analysis, through human assistance, artificial intelligence or a rare combination of both, and application of the robots capabilities, through programming, for tasks required based on the analysis. Thus, Robotic devices are primarily smart machines with manipulators that can be programmed to do a variety of manual or human labor tasks automatically, and with sensors that explore the surrounding environment.

Space robots share certain common features with their terrestrial counterparts. They also involve, however, a blending of aerospace and computer technologies that is far more demanding, unusual, and sophisticated than that generally needed for robots operating on Earth. Space robots have to work in the harsh environment of outer space and sometimes up on strange alien worlds about which little is previously known. Space robots need to be smart, i.e., they need to possess a certain level of artificial intelligence (A.I.) to perform tasks dedicated to it. The need for robots to possess A.I. arises due to the limitations of communication between human controllers on Earth and the space robots. Under certain circumstances, telepresence and virtual reality technologies will allow a human being to form a real-time, interactive partnership with an advanced space robot, which serves as a dexterous mechanical surrogate capable of operating in a hazardous, alien world environment.

As a space robot operates farther away, the round-trip communications distance with human controllers back on Earth must soon be measured not in thousands of miles (or kilometers), but rather in light minutes. The great distances associated with deep-space exploration make the real-time control of a robot spacecraft by human managers impractical. To maintain continuous communication with the space robot also becomes impossible under certain circumstances because of the loss of line of sight between the robot and signal receivers on Earth. For example, if the robot landed on the side of a planet hidden to Earth, it becomes impossible to send or receive signals from the robot. Therefore, in order to survive and function around or on distant worlds, they need to contain various levels of machine intelligence, or artificial intelligence (AI).

1.1.   The need for robotics in space
Modern space robots are sophisticated exploring machines that have, or will have, visited all the major worlds of the solar system, including

(soon) tiny Pluto. Space is a hostile environment, which is not suited for the existence of human beings. The only other celestial body that human beings have set foot on besides the Earth is moon. Robotic space probes, however, have done most of the exploration on worlds such as Mars, Venus, Saturn, Jupiter and a few comets and asteroids.

It is important simply to recognize that sophisticated robot spacecraft represent the enabling technology for many of the most important scientific discoveries that await the human race in the remainder of this century. Space robots are the mechanical partners that enable the human race to fulfill its destiny as an intelligent, spacefaring species. Failure to fully appreciate or to capitalize upon the opportunity offered by the space robot will confine future generations of human beings to life on just one planet around an average star in the outer regions of the Milky Way Galaxy. By recognizing the value of and vigorously using the space robot, the human race will, however, emerge within the galaxy as an active, spacefaring species. By initially reaching for the stars with very smart machines, future generations of human beings will experience all of the exciting social and technical impacts involved in becoming an interstellar spacefaring species.

Human designed space robots have developed from unsophisticated unreliable machines to sophisticated efficient machines due to the advance of computer and aerospace technologies in the recent years. We have become more familiar to our solar system in the past 50 years or so than the human race had been able to do since the remainder of the history of Earth. The first family of space robots, which were given the name *Pioneer*, were the first space missions to be carried out by the United States Air Force in 1958. These early Pioneer lunar probes were the world's first attempted deep-space missions. The first series of Pioneer spacecraft was flown between 1958 and 1960. *Pioneer 1, 2,* and *5* were developed by Space Technology Laboratories, Inc. and were launched for NASA by the Air Force Ballistic Missile Division (AFBMD). *Pioneer 3* and *4* were developed by the Jet Propulsion Laboratory (JPL) and launched for NASA by the U.S. Army Ballistic Missile Agency (ABMA). These were some of the early attempts. As compared to these partially successful easy and unsophisticated missions, space robots efficiency and sophistication was improved to perform successful missions such as *Mariner* 2, which was the first robot spacecraft to fly past another planet (Venus), and *Mariner* 4 which voyaged the red planet, Mars. In 1977, NASA launched two sophisticated space robots, *Voyager* 1 and 2. *Voyager 1* visited Jupiter and Saturn, while *Voyager 2* took the so-called "grand tour" and visited all four giant planets on the same mission. One of the most successful missions, the Galileo mission began on October 18, 1989, when the sophisticated spacecraft was carried into low Earth orbit by the space shuttle *Atlantis* and then launched on its interplanetary journey by means of an inertial upper stage (IUS) rocket. Relying on gravity assist flybys, to reach Jupiter, the *Galileo* spacecraft flew past Venus once and

Earth twice. As it traveled through interplanetary space beyond Mars on its way to Jupiter. *Galileo* started a second extended scientific mission in early 2000. This second extended mission included flybys of the Galilean moons Io, Ganymede, and Callisto, plus coordinated observations of Jupiter with the *Cassini* spacecraft. In December 2000, *Cassini* flew past the giant planet to receive a much-needed gravity assist that enabled the large spacecraft to eventually reach Saturn. *Galileo* conducted its final flyby of a Jovian moon in November 2002, when it zipped past the tiny inner moon, Amalthea.

Chapter 2: SPACE ROBOT SPECIFICS

A space robot is an unmanned platform, designed to be placed in an orbit around earth or an interplanetary trajectory to another celestial body. The space robot is essentially a combination of hardware that forms such a mission-oriented spacecraft. A robotic spacecraft consists of four parts. The hierarchy of aerospace hardware is system, subsystem, assembly, and component (or part) in that descending order. A robot spacecraft is very complex in context to the interdependent nature of many of its systems and subsystems. Thus, the application of this hardware classification scheme is very arbitrary in such spacecrafts. Fortunately, a little apparent confusion in nomenclature in no way detracts from the quality of the hardware. Individual robot spacecraft can be very different from one another in design and level of complexity, including the type and number of subsystems and component parts and assemblies found in each individual subsystem. For example, a robot probe, which descends into a planetary atmosphere on a one-way scientific mission, will generally not have a propulsion subsystem or an attitude-control subsystem.

2.1. Machine Intelligence

Different space robots possess different levels of machine intelligence. A robot's level of machine intelligence determines the degree of autonomous operation possible and the amount of human supervision required. For deep space missions, direct human supervision is usually impractical or impossible. Therefore, a space robot engaged in this type of mission must have an appreciable level of machine intelligence.

The term artificial intelligence (AI) is commonly referred to as the study of thinking and perceiving as general information-processing functions. Also referred to as, the science of machine intelligence (MI). Perception is the process of obtaining data from one or more sensors and processing or analyzing these data. This analysis assists in making some subsequent decision. The basic problem in perception is to collect a large amount of (remotely) sensed data and extract from it and analyze the data that then permits object identification. All artificial intelligence involves elements of planning and problem solving. The problem solving function implies a wide range of tasks, including decision making, optimization, dynamic resource allocation, and many other calculations or logical operations. This analysis is put into action through mechanical subsystem devices, such as end effectors, controlled by the robot itself.

Robots may be mobile or fixed in place and either fully automatic or teleoperated. The more AI a robot has the less dependent it is upon human supervision. As levels of machine intelligence continue to improve in this century, truly autonomous space robots will become a reality. Artificial intelligence experts suggest that smart exploring machines of the future will have (machine) intelligence capabilities sufficient to repair themselves, to avoid hazardous circumstances on alien worlds, and to recognize and report all of the interesting objects or phenomena they encounter.

2.2 General Classification of Scientific Spacecraft

Since different space missions require customized and optimized space robots, it is necessary to obtain information about their classes in order to obtain knowledge about their subsystem assemblies.

Trajectory Classification

There are three basic possibilities for a robot spacecraft's trajectory when it encounters a planet:

1. The first discussed trajectories are for missions that require descending of the lander spacecraft to the surface of a planet or a moon.

    The first possible trajectory involves a direct hit or hard landing. This is an impact trajectory. A hard landing involves a relatively high-velocity impact landing of the robot spacecraft on the surface of a planet or moon. This usually destroys all equipment, except perhaps for a very rugged instrument package or payload container. In the second possibility, the lander spacecraft follows a specific trajectory to the surface of another planet and lands softly onto the surface of the planet, for safety of the robot and the equipment it contains. This can be achieved either by sending the lander spacecraft on a direct trajectory from the earth to the surface of the planet, or by a soft release into a specific trajectory by a mother spacecraft. The soft release can be sequenced after the mother spacecraft has entered into orbiting around the particular planetary body and located a vantage point from which to execute the release. In another lander/probe mission scenario, the mother spacecraft releases the lander or robot probe, while the co-joined spacecraft pair is still some distance from the target planetary object. Following release and separation, the robot probe follows a ballistic impact trajectory into the atmosphere and onto the surface of the target body. The mother spacecraft can change its trajectory from then onwards by firing thrusters and continue on mission somewhere else.

2. The second type of trajectory is an orbital-capture trajectory. The spacecraft is captured by the gravitational field of the planet and enters orbit around it. Depending upon its precise speed and altitude (and other parameters), the robot spacecraft can enter this captured orbit from either the trailing edge or the leading edge of the planet.

3. In the third type of trajectory, called a flyby trajectory, the spacecraft remains far enough away from the planet to avoid capture, but passes close enough to be strongly affected by its gravity. In this case, the speed of the spacecraft will be increased, if it approaches from the trailing side of the planet, and diminished if it approaches from the leading side. In addition to changing speed, the spacecraft's motion also changes direction. The increase in speed of the flyby spacecraft actually comes from a decrease in speed of the planet itself. In effect, the spacecraft is being "pulled along" by the planet i.e. it is travelling the extra path that the planet is travelling relative to the sun and being pulled along due to its gravity, thus increasing the spacecrafts velocity relative to the sun. The increase in momentum of the spacecraft results from the decrease in momentum of the planet. However, due to the extreme difference in their masses, the decrement in velocity of the planet is negligible. A full account of spacecraft trajectories must consider the speed and actual trajectory of the spacecraft and planet, how close the spacecraft will come to the planet, and the size (mass) and orbital speed of the planet, in order to make even a simple calculation.

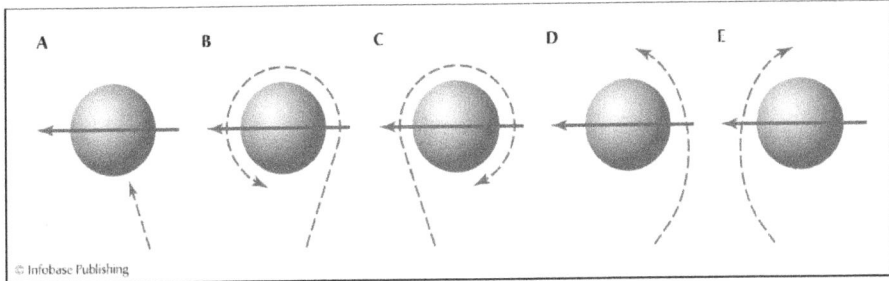

Figure 1.1: Different planetary flybys

The figure above illustrates the possible trajectories of a spacecraft encountering a planet. Case A shows a direct hit or hard landing. Cases B and C show orbital capture trajectory from trailing edge and leading edge respectively. Case D and E show flybys from the trailing and leading edge respectively.

Based on the classification given above and mission requirements, a robot spacecraft can be: flyby spacecraft, orbiter spacecraft, atmospheric probe spacecraft, atmospheric balloon packages, lander spacecraft, surface penetrator spacecraft, surface rover spacecraft, and observatory spacecraft. These diverse types of spacecraft possess different structure and functional subsystems, which are mission compatible.

Flyby spacecrafts follows a continuous trajectory through space and are not captured by the planet into planetary orbit. These spacecrafts are designed to observe passing celestial targets through their equipment.

Such encounters of the spacecraft with celestial objects last for a few hours or minutes. Thus, they must possess the capability of travelling on powered down, cruise mode for years and bringing all their senses into focus rapidly. They must also be capable of high rates of data transmission to Earth and storing this data when the antenna is not pointing towards Earth. NASA's *Pioneer 10* and *11* and the *Voyager 1* and *2* are examples of highly successful flyby scientific spacecraft.

An orbiter spacecraft is designed to travel to a distant planet and then orbit around that planet. This type of scientific spacecraft must possess a substantial propulsive capability to decelerate at just the right moment in order to achieve a proper orbit insertion. Solar occultations have to be dealt with while designing an orbiter spacecraft. These occultations causes cut off from solar powered electricity generation, extreme variations in thermal environment of the spacecraft and inability to constantly transmit data to Earth. Thus, generally, a rechargeable battery system augments solar electric power. Active thermal control techniques are used to complement traditional passive thermal-control design features. Interruption in uplink and downlink communications with Earth, make onboard data storage necessary. The *Lunar Orbiter, Magellan, Galileo,* and *Cassini* are examples of successful scientific orbiters.

Atmospheric probe spacecraft are small, instrumented spacecraft, which study the gaseous atmosphere of a planet or a moon while they descend through it. These probes are separated from their mother spacecraft prior to its closest approach to the planet. The mother spacecraft makes corrections in trajectory to escape the planets gravity and carry on other missions. An aeroshell usually protects these atmospheric probes from the atmospheric friction caused during entry. A parachute is deployed to slowly descend the spaceprobe through the atmosphere while scientific instruments onboard study the surrounding environment. The data is telemetered from the atmospheric spaceprobe to the mother spacecraft, which then relays the data to Earth. NASA's *Pioneer Venus* (four probes), *Galileo* (one probe), and *Cassini* (*Huygens probe)* missions involved the deployment of a probe or probes into the target planetary body's atmosphere (i.e., *Venus, Jupiter*, and Saturn's moon *Titan*, respectively).

An atmospheric balloon package is designed for suspension from a buoyant gas filled bag that can float and travel under the influence of the winds in a planetary atmosphere. Tracking of the balloon package's progress across the face of the target planet will yield data about the general circulation patterns of the planet's atmosphere. A balloon package needs a power supply and a telecommunications system. It also can be equipped with a variety of scientific instruments to measure the planetary atmosphere's composition, temperature, pressure, and density. During their flyby of Venus in June 1985, the Russian *Vega 1* and *2* deployed constant pressure instrumented balloon aerostats. Data (such as temperature, pressure, and wind velocity) from each balloon's scientific instruments was transmitted directly to Earth for the 47-hour lifetime of the aerostat mission.

Lander spacecrafts are designed to reach a planet's surface and send useful information such as imagery of landing site, measurement of local environment conditions, and initial examination of soil composition. For example, NASA's Surveyor lander craft extensively explored the lunar surface at several landing sites in preparation for the human Apollo Project landing missions, while NASA's *Viking 1* and *2* lander craft investigated the surface conditions of Mars at two separate sites for many months.

A surface penetrator is a robot spacecraft, designed to enter the solid surface of a planet, an asteroid or comet and transmit back to the mother spacecraft the relevant data. It must be capable of surviving a high velocity hard hit or direct landing onto the surface of the planet. A surface rover spacecraft, on the other hand, is soft landed on the surface of a planet via a specific trajectory from the mother spacecraft. Once deployed on the surface, the electrically powered rover can wander a certain distance away from the landing site, take images and perform soil analyses. Data then is telemetered back to Earth by one of several techniques: via the lander spacecraft, via an orbiting mother spacecraft, or (depending on size of rover) directly from the rover vehicle. The Soviet Union deployed two highly successful robot surface rovers called *Lunokhod 1* and *2* on the Moon in the 1970s. In December 1996, NASA launched the Mars Pathfinder mission to the Red Planet. A successful mission, the pathfinder collected and transmitted data, from its innovative airbag-protected bounce and role landing on July 4, 1997, until the final data transmission on September 27. The lander/rover team returned numerous close-up images of Mars and chemical analysis of various rocks and soil found near the landing site.

An observatory spacecraft is a space robot that does not travel to a destination to explore. Instead, this type of robot spacecraft travels in an orbit around Earth or around the Sun. From vantage points in its orbit, the observatory can view distant celestial targets unhindered by the blurring and obscuring effects of Earth's atmosphere. NASA's Great Observatories Program consists of a family of four orbiting observatories each studying the universe in a different portion of the electromagnetic spectrum. Spitzer Space telescope is the final mission of the Great observatory program. It represents the most powerful and sensitive infrared telescope ever launched. The orbiting facility obtains images and spectra of celestial objects at infrared radiation wavelengths between 3 and 180 micrometers (μm)—an important spectral region of observation mostly unavailable to ground-based telescopes because of the blocking influence of Earth's atmosphere. Other missions in this program include the *Hubble Space Telescope (HST),* the *Compton*

Gamma Ray Observatory (CGRO), and the *Chandra X-ray Observatory (CXO).*

2.3. Functional Subsystems

A robot spacecraft's functional subsystems support the mission oriented science payload and allow the spacecraft to operate in space, collect data and transmit it back to earth. There are certain general subsystems, which are necessary for almost all missions while certain subsystems are mission specific. These subsystems work mutually to carry on a mission successfully. They are interdependent and the whole mission is designed around the abilities of these subsystems. Some of the common subsystems are, as explained below:

2.3.1. Structural subsystem

The structural subsystem supports all other subsystems as well as the payload of scientific instruments. All of the other spacecraft components are attached to the structural subsystem. Aluminium is by far the most common spacecraft structural material. The engineer can select from a wide variety of aluminium alloys, which provide the spacecraft designer with a broad range of physical characteristics, such as strength and machinability. A space robot's structure may also contain magnesium, titanium, beryllium, steel, fiberglass, or low mass and high-strength carbon composite materials. Of course, the design of the structural subsystem, for each space robot is unique. It demands the complete knowledge of mission requirements and the other components that will be attached to the structural subsystem. For example, a hard hit landed rover will require a much greater strength and resistance that a soft landed space probe, to prevent its subsystems from damage during crash. The environment of the planet to which the space robot has to encounter also plays an important role in determining the structural characteristics of the space robot.

2.3.2. Power subsystem

The power subsystem installed in a space robot again depends upon the mission requirements of the probe, how long is it supposed to transmit data to earth and the functions it must perform during the course of its exploration. On an average, a complex robot spacecraft needs between 300 and 3,000 watts to properly conduct its mission. The less power available, however, means the less performance and flexibility the engineers can give the space robot. Small short lived robot spacecraft, such as an atmospheric probe and a mini-rover, might need only 25 to 100 watts, which can often be supplied by long-lived batteries. Since, the power subsystem must satisfy all the need of the robot spaceprobe , its construction varies with variation in mission requirements and conditions.

Space probes operating within the orbit of Mercury and Mars are best powered by use of a solar-photovoltaic (solar-cell) system, in combination with rechargeable batteries, to provide a continuous supply of electricity. The spacecraft must also have a well-designed, built-in

27

electric utility grid, which conditions and distributes power to all onboard components. Solar cells do not work well on spacecraft that must fly very close to the Sun, because of the severe thermal environment encountered. For space probes travelling beyond the orbit of Mars, Solar-photovoltaic power subsystem becomes impractical. Moreover, for missions into deep space or in hostile planetary environment, it becomes almost impossible to utilize. Under these mission circumstances, the use of a long-lived nuclear power supply becomes practical. This type of power subsystem is called a radioisotope thermoelectric generator (RTG). The RTG converts the decay heat from a radioisotope directly into electricity by means of the thermoelectric effect.

2.3.3. Attitude control subsystem

A spacecraft's attitude-control subsystem includes the onboard system of computers, low-thrust rockets (thrusters), and mechanical devices (such as a momentum wheel) used to keep a spacecraft stabilized during flight and to precisely point its instruments in some desired direction. *Stabilization* refers to maintaining the spacecrafts orientation in space, relative to Earth, as required by the command control. It is achieved by either spinning the spacecraft or by using a three-axis active approach that maintains the spacecraft in a fixed, reference attitude by having it fire a selected combination of thrusters when necessary.

Stabilization achieved by spinning the spacecraft uses the gyroscopic action of the rotating mass to keep the spacecraft stabilized; as was done in *pioneer 10* and *11*. Active three-axis stabilization can be achieved either by the use a small propulsive subsystem or through reaction wheels. The small propulsive subsystem consists of a number of thrusters placed around the probe. These thrusters, fired in proper combinations, nudge the space probe back and forth within the allowed band of altitude error. The other method, using electrically powered reaction wheels, also known as momentum wheels, utilizes Newton's third law. These massive wheels are mounted in three orthogonal axes onboard the spacecraft. To rotate the spacecraft in one direction, the proper wheel is spun in the opposite direction. To rotate the vehicle back, the wheel is slowed down. Excessive momentum, which builds up in the system due to internal friction and external forces, occasionally must be removed from the system; this usually is accomplished with propulsive maneuvers. Both the general approaches to spacecraft stabilization have their basic advantages and disadvantages. Spin-stabilized vehicles provide a continuous "sweeping motion" that is generally desirable for fields and particle instruments. Such spacecraft, however, may then require complicated systems to de-spin antennae or optical instruments that must be pointed at targets in space. Three-axis controlled spacecraft can point antennae and optical instruments precisely, but these craft then may have to perform rotation maneuvers to use their field and particle science instruments properly.

An important function of the altitude control subsystem is to work very closely in conjunction with the propulsive subsystem. The altitude control subsystem makes sure the space probe is pointing in the right direction before a major rocket engine burn. In addition, the process of orbit insertion requires precise positioning and controlled deceleration. Here as well, the altitude control subsystem works in coordination with the propulsive subsystem.

2.3.4. Thermal control subsystem

The thermal-control subsystem regulates the temperature of a robot spacecraft and keeps it from getting too hot or too cool. Thermal control is a complex problem because of the severe temperature extremes a space robot experiences during a typical scientific mission. The overall thermal energy balance for a spacecraft near a planetary body is determined by several factors: thermal energy sources within the spacecraft, direct solar radiation, direct thermal (infrared) radiation from the planet, indirect (reflected) solar radiation from the planetary body, and thermal radiation emitted from the surface of spacecraft to the low temperature sink of outer space. In the vacuum environment of outer space, radiation-heat transfer is the only natural mechanism for exchanging thermal energy (heat) into or out of a spacecraft, whereas, conduction-heat transfer generally controls the flow of heat within the spacecraft. During the periods of shadow, i.e. no access to Sun's infrared radiation, temperature of the space probe orbiting the Earth can drop to 200 K and can rise upto 350 K during periods encountering infrared radiation from sun. Spacecraft materials and components can experience thermal fatigue due to repeated extreme temperature cycles. Thus, passive and active thermal control techniques are used to regulate the temperature of spacecraft components. Passive thermal control techniques include the use of special paints and coatings, insulation blankets, radiating fins, sun shields, heat pipes, as well as careful selection of the spacecraft's overall geometry. Active thermal control techniques include the use of heaters (including small radioisotope sources) and coolers, louvers and shutters, or the closed-loop pumping of cryogenic materials. The small radioisotope heat sources are especially useful for specific components on lander and rover robots that must stay within certain temperature limits in order to survive the frigid nighttime conditions experienced on the surface of the Moon or Mars.

2.4. Telecommunications

Telecommunications refers to the flow of data and information (usually by radio signals) between a spacecraft and an Earth-based communications system. The radio signals must travel millions, sometimes even billion (in deep space missions) of kilometers, travelling from the space probe to the Earth-based communications system. Due to the limit amount of power supply, a space probe has no more than 20 Watts of radiating power. In order to conduct telecommunications effectively, all the radio signal power available are concentrated down into a narrow radio beam, using a parabolic dish antenna of the order three to 15 feet (1 to 5 m) in diameter. This beam in sent to one specific direction (Earth), instead of broadcasting it in all directions. Even when these concentrated radio signals reach Earth, however, they have very small power levels. Therefore, special, large-diameter radio receivers on Earth, such as found in NASA's Deep Space Network are used. These sophisticated radio antennae are capable of detecting the very-low-power signals from distant spacecrafts.

In telecommunications, the radio signal transmitted to a spacecraft is called the uplink. The transmission from the spacecraft to Earth is called the downlink. The spacecraft's carrier signals are modulated by shifting each waveform's phase slightly at a given rate. To modulate the carrier with a frequency, for example, one megahertz (MHz), the frequency is modulated to carry individual phase shifts that are designated to represent binary ones (1s) and zeros (0s). This is the spacecraft's telemetry data. The amount of phase shift used in modulating data onto the subcarrier is referred to as the modulation index and is measured in degrees. This same type of communications scheme is also on the uplink.

Two types of antenna are used on a space probe to transmit and receive communications from Earth: *High Gain Antenna (HGA)* and *Low Gain Antenna (LGA)*. The amount of gain achieved by an antenna refers to the amount of incoming radio signal power it can collect and focus into the spacecraft's receivers. The high-gain antenna incorporates a large parabolic reflector. The larger collecting area of the high-gain antenna supports higher data rate. But, higher the gain, more directional the antenna becomes. Thus, when communications take place with the use of an *HGA*, the antenna has to be pointed within a fraction of a degree of Earth in order for communications to take place. An *LGA*, on the other hand supports a much low data rate, but it has a wide-angle coverage. It is useful as long as the spacecraft is close to earth or when the *HGA* has Earth covered by a blind spot.

Chapter 3: Flyby Spacecrafts

Since our aim is to calculate the trajectory of a scientific spacecraft from planet Earth to planet Neptune, we shall take a look at some of the important flyby trajectories that have changed the course of space propulsion greatly. A review on the planetary flyby missions will help gain a wider perspective on the mechanics of bodies under the force of gravity alone. In order to reach distances such as the edge of the solar system, as in this case, Neptune, it is essential that the spacecraft is sent on its course to the target planet via as many flyby's as possible. A planetary flyby maneuver can be utilized as a means to increase velocity of the spacecraft relative to the Sun. This allows the spacecraft to save a considerable amount of time to reach its destination.

A flyby mission is an interplanetary or deep space mission in which the robot spacecraft passes close to its designated celestial target, but does not impact the target or go into orbit around it. Flyby spacecraft follow a continuous trajectory to that allows them to not be captured into a planetary orbit but passes close enough to be strongly affected by the celestial bodies gravitation. Once the spacecraft has flown past its target, it cannot return to recover lost data as it isn't in an orbit around the target. Therefore, flyby operations often planned years prior to launch and practiced to perfect the gravity assist maneuvers that the flyby spacecraft will have to make use of during the course of the mission.

During a flyby's trajectory, the spacecraft remains far enough away from the planet to avoid capture, but passes close enough to be strongly affected by its gravity. In this case, the speed of the spacecraft increases if it approaches from the trailing side of the planet and diminished if it approaches from the leading side. During a flyby, the spacecraft's motion also changes direction. The increase in speed of the flyby spacecraft actually comes from a decrease in speed of the planet itself. In effect, the spacecraft is being "pulled along" by the planet. This "pulling along" phenomenon, which changes the velocity and direction of motion of the spacecraft, is known as a *gravity-assist* maneuver.

3.1. Gravity assist (or the mechanics of a flyby)

A gravity assist or slingshot maneuver around a planet changes a spacecraft's velocity relative to the Sun, even though it preserves the spacecraft's speed relative to the planet—as it must according to the law of conservation of energy. This an elastic collision even though no actual contact occurs. Gravity assist can be explained with the use of vectors in a slightly detailed mathematical description based on figure 2.

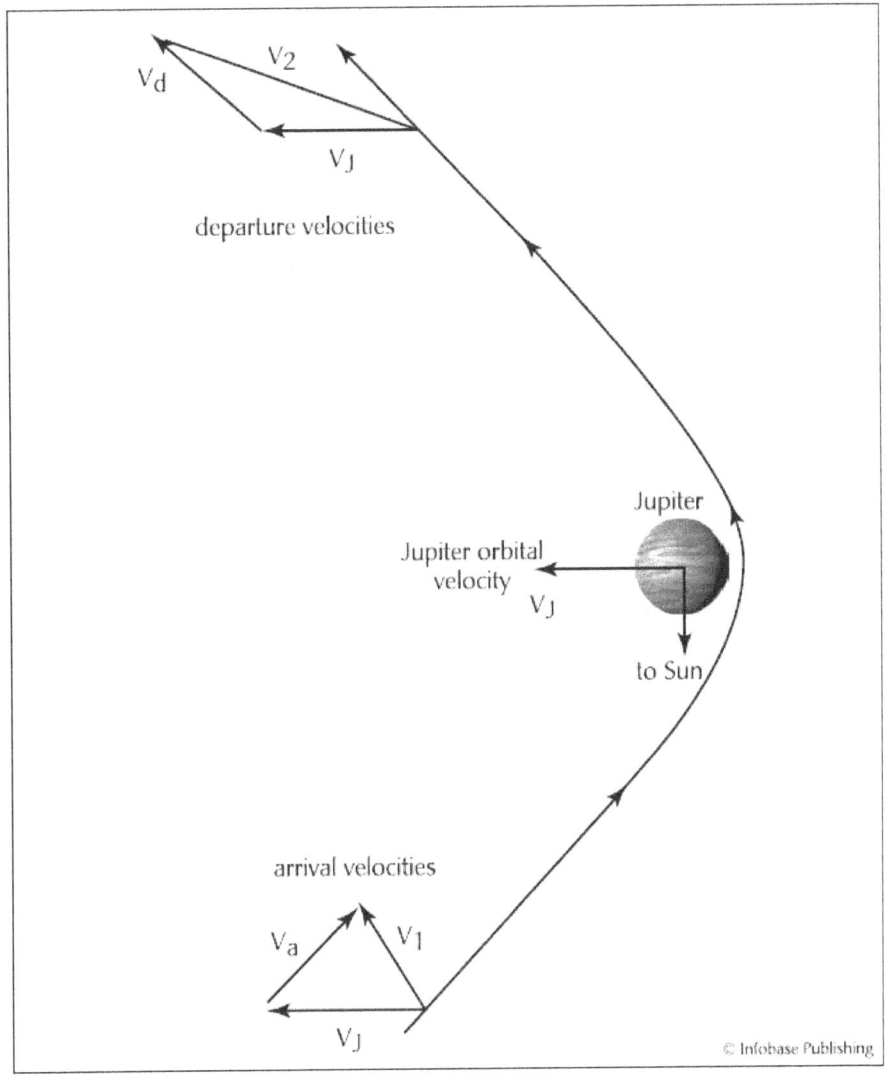

Figure 3.1: Gravity Assist

Voyager 1 and *2* spacecrafts performed gravity assist on Jupiter to reach Saturn and save time and energy. the heliocentric (Sun-centered) path each spacecraft followed in its motion with respect to Jupiter was closely approximated by a hyperbola.

The approach velocity (V_1) of the spacecraft is the vector sum of the orbital velocity of Jupiter (V_j) and the velocity of the spacecraft with respect to Jupiter. The spacecraft moves toward Jupiter along an asymptote, approaching from the approximate direction of the Sun and with velocity relative to Jupiter Va. The velocity relative to Sun (V_1) is then computed by vector addition: $V_1 = V_j + V_a$. The spacecraft then departs Jupiter in a new direction, determined by the amount of bending that is caused by the effects of the gravitational attraction of Jupiter's mass upon the mass of the spacecraft. The departure speed relative to Jupiter (V_d) on the hyperbola is equal to the arrival speed. Thus, the length of V_a equals the length of V_d. For the departure relative to Sun, the velocity is: $V2 = V_j + V_d$ .During the relatively short period of time that the spacecraft is near Jupiter, the orbital velocity of Jupiter (V_j) changes very little, and so V_j is assumed to be a constant.

The vector sums in the figure shows that the deflection, or bending, of the spacecraft's trajectory caused by Jupiter's gravity results in an increase in the speed of the spacecraft *along its hyperbolic path,* as measured relative to the Sun.

3.2. Phases of a Flyby Mission

Flyby operations are conveniently divided into four phases: observatory phase, far-encounter phase, near-encounter phase, and post-encounter phase.

3.2.1. Observatory Phase

The observatory phase generally begins a few months prior to the date of the actual planetary flyby when the spacecraft is far from the celestial target and the planet can fully fit within the field of view of the spacecrafts instruments. During this period, the celestial target can be better resolved in the spacecraft's optical instruments than it can from Earth-based instruments.   During this phase, the instruments of the spacecraft become completely operational in support of the forthcoming encounter, and the spacecraft is totally involved in making observations of its target.

3.2.2. Far Encounter Phase

The far-encounter phase includes time during the approach of the spacecraft when the full disk of the target planet (or other celestial object) can no longer fit within the field of view of the spacecraft's instruments. Taking advantage of the higher resolution available when the spacecraft is close to the planetary body, observations during this phase are designed to focus on parts of the planetary body (for example, the Caloris Crater on Mercury, Jupiter's Red Spot, or the cantaloupe-like surface features on Triton), rather than the entire planet.

3.3.3. Near Encounter Phase:

The near-encounter phase is the period of closest approach of the spacecraft to the target. It is characterized by intensely active observations of the target body by all of the spacecraft's science experiments. This closest phase of the flyby spacecrafts encounter with the target provides scientists the opportunity to obtain the highest resolution data about the target. Detailed observations are made during this phase of observation with fully active resources and scientific equipment. This phase provides data essential for detailed study on the planetary object. It is essential that during the near phase encounter, the flyby spacecraft collects data from all the equipment that wouldn't provide better results from far off.

3.3.4. Post Encounter Phase

The post-encounter phase begins when the near-encounter phase is completed and the spacecraft is receding from the target. This phase is characterized by day-after-day observations of a diminishing planet just encountered. While receding from the planet, the spacecraft is likely to observe the night side of the planet. This provides an opportunity to make extensive observations of the night side of the planet.

After the post-encounter phase is over, the spacecraft stops observing the target and returns to the less intense activities of its interplanetary cruise phase. Cruise phase can be defined as a phase in which scientific instruments are powered down and navigational corrections made to prepare the spacecraft for an encounter with another celestial object of opportunity or for a final journey of no return into deep space. Some scientific experiments, usually concerning the properties of interplanetary space, can be performed in this cruise phase.

To understand the mission requirements of a flyby spacecraft better, we shall discuss three of NASA's important flyby missions: *Mariner 10* mission to Mercury, by way of a gravity-assist from Venus; *Pioneer 11*'s flyby encounter of Saturn, following its earlier close encounter with Jupiter, which provided an important gravity-assist; and, the *Voyager 2* spacecraft whose grand-tour mission took the hardy space robot past all of the giant planets.

3.4. Important Flyby Missions
In this section contains a review of three of important flyby missions launched by NASA. These missions were pioneer missions that implemented the use of new techniques in gravity assist maneuvers. They made deep space explorations possible with the use of gravity assist maneuvers. Voyager 2 illustrated a breakthrough in gravity assist maneuver as it visited the four giant planets with the help of repeated gravity assist.

3.4.1. Mariner 10

The Mariner 10 spacecraft was the first to use the gravitational pull of one planet (Venus) to reach another planet (Mercury). It passed Venus on February 5, 1974, at a distance of 4,200 km. The robot spacecraft then crossed the orbit of Mercury at a distance of 704 km from the surface on March 29, 1974. A second encounter with Mercury occurred on September 21, 1974, at an altitude of about 47,000 km from the surface of Mercury. A third and final Mercury encounter took place on March 16, 1975, when the spacecraft passed the planet at an altitude of 327 km from Mercury. When the supply of attitude-control gas became depleted on March 24, 1975, this highly successful mission was terminated.

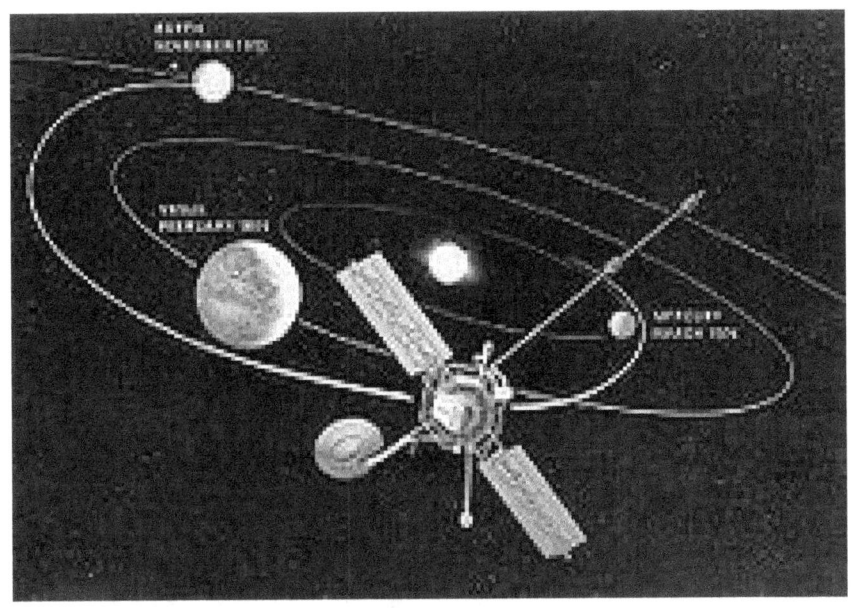

Figure 3.2: Mariner 10

The Mariner 10 spacecraft structure was an eight-sided magnesium frame with eight electronics compartments. The space robot measured 1.39 m diagonally and 0.46 m in depth. Two solar panels, each 2.69 m long and 0.97 m wide, were attached at the top, supporting 5.1 square meters of solar-cell area. Fully deployed Mariner 10 measured 8 m across the solar panels and 3.7 m from the top of the low-gain antenna to the bottom of the heat shield.

Figure 3.3: Another look at the power source

Engineers mounted a scan platform with two degrees of freedom on the anti-solar face of the spacecraft structure. The spacecraft had a total launch mass of 503 kg, of which 29 kg were propellant and attitude-control gas. The total mass of onboard instruments was 79 kg. The spacecraft carried science instruments to measure the atmospheric, surface, and physical characteristics of Venus and Mercury. Experiments included television-photography, infrared radiometers, and ultraviolet spectroscopy.

The Mariner 10's rocket engine was a 50-pound-force (222-newton) liquid monopropellant hydrazine motor located below a spherical propellant tank, which was mounted in the center of the structural framework. The rocket nozzle protruded through a sunshade. Engineers used a total of six (two sets of three orthogonal pairs) pressurized nitrogen gas reaction- jets (thrusters), which they mounted on the tips of the solar panels to achieve three-axis stabilization of the spacecraft. Command and control of these thrusters were the responsibility of an on-board computer.

Finally for communications, Mariner 10 carried a motor-driven high-gain dish antenna, with a 1.37-m diameter parabolic reflector made of aluminum honeycomb sandwich material. This high-gain antenna was mounted on a boom on the side of the spacecraft. The spacecraft also had a low-gain, omnidirectional antenna, which was mounted at the end of a 2.85-m long boom, extending from the anti-solar face of the spacecraft. Mariner 10 was the first and, thus far, the only spacecraft of any country to explore the innermost planet in the solar system.

3.4.2. Pioneer 10 and 11
NASA's Pioneer 11 spacecraft and Pioneer 10 were the first spacecrafts to navigate the main asteroid belt. Pioneer 10 was the first spacecraft to encounter Jupiter and its fierce radiation belts and the first human-made object to leave the planetary boundary of the solar system.

Pioneer 11 the first spacecraft to encounter Saturn. Pioneer 11 was backup for Pioneer 10 incase of failure. But it was also reprogrammable for a trajectory to Saturn using gravity assist from Jupiter. Its trajectory was retargeted to "slingshot around Jupiter and reach Saturn. The

Pioneer 11 spacecraft was launched on April 5, 1973, and swept by Jupiter at an encounter distance of only 43,000 km on December 2, 1974. Then, on September 1, 1979, Pioneer 11 flew by Saturn, demonstrating a safe flight path through the rings for the more sophisticated Voyager 1 and 2 spacecraft to follow. Pioneer 11 provided the first close-up observations of Saturn, its rings, satellites, magnetic field, radiation belts, and atmosphere.

As they flew through interplanetary space, these spacecraft also investigated magnetic fields, cosmic rays, solar wind, and interplanetary dust concentrations. Pioneer 10 and 11 were identical in construction. The Pioneer 11 spacecraft contained the following distinct subsystems: a general structure, an attitude control and propulsion system, a communications system, thermal control system, electric power system, navigation system, and a science payload (containing 11 instruments).

Basically, the Pioneer 11 (and its Pioneer 10 technical twin) had to be extremely reliable and lightweight. The spacecraft needed a communications system capable of transmitting data over extremely large distances. Since each spacecraft would operate so far from the Sun, engineers chose a nuclear (non-solar) power source for electric power generation.

Pioneer 11 was 2.9 m long. It contained a 2.74 m diameter, 46 cm deep, parabolic, dish-shaped high-gain antenna made of aluminum honeycomb sandwich material to the front of the equipment compartment. The feed of the high-gain antenna was coupled with a medium-gain antenna mounted on three struts, and a 0.76-m diameter lowgain, omnidirectional antenna below the dish of the high-gain antenna.

The robot spacecraft had three reference sensors to support interplanetary navigation: a star (Canopus) sensor and two Sun sensors. Attitude position could be calculated from the reference direction to Earth and the Sun, with the known direction to the star Canopus used as a backup.

Pioneer 11 had three pairs of rocket thrusters, which could be fired on command either steadily or in pulses. Three pairs of rocket thrusters located near the rim of the antenna dish were used to direct the spin axis of the spacecraft, to keep the spacecraft spinning at the desired rate of 4.8 revolutions per minute (rpm), and to change the velocity of the spacecraft for in-flight

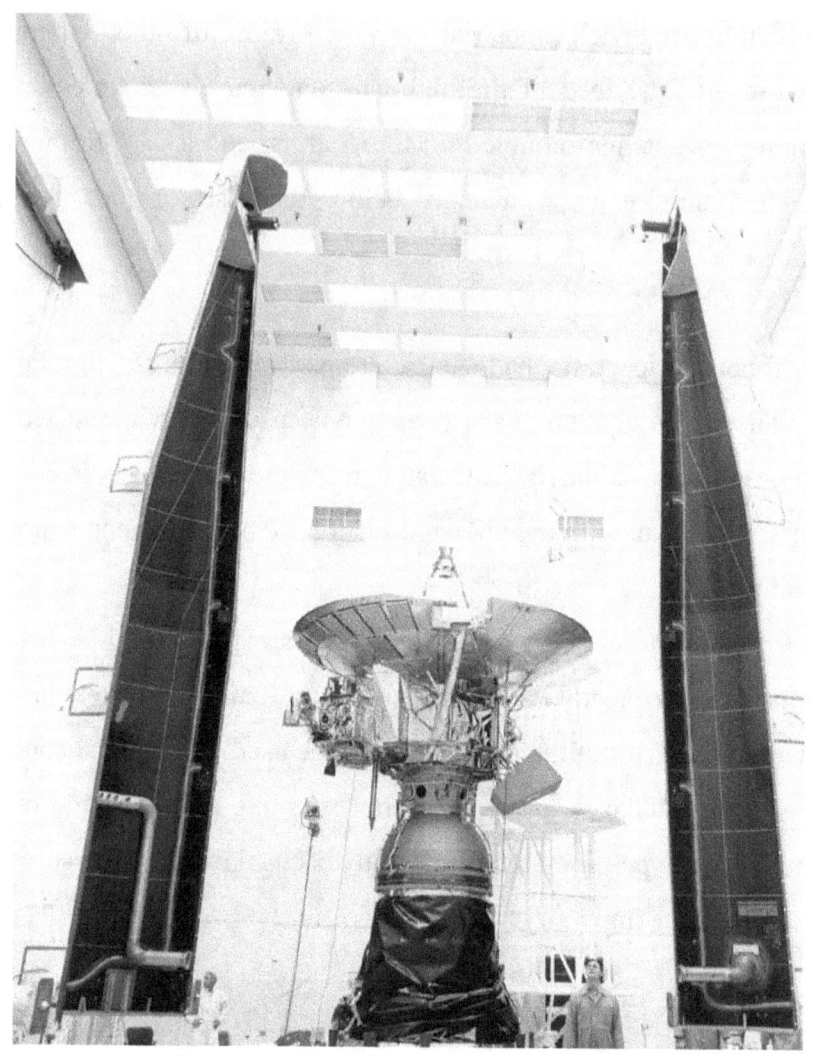

Figure 3.4: Pioneer 11

maneuvers. The spacecraft's six thrusters could be commanded to fire steadily or in pulses. Each thruster developed its propulsive jet from the decomposition of liquid hydrazine by a catalyst in a small rocket thrust chamber to which the nozzles of the thruster were attached.

The spacecraft's thermal-control system was designed to maintain the temperature inside the science instrument compartment between -23°C and +38°C. One specific feature of pioneer 11 is that, the thermal control system was designed to adapt to the gradual decrease in solar heating as *Pioneer 11* moved away from the Sun. It was also constructed to survive periods when the spacecraft passed through Earth's shadow (after launch), followed by Jupiter's shadow, and then Saturn's shadow, during the planetary encounters.

3.4.3. Voyager 2

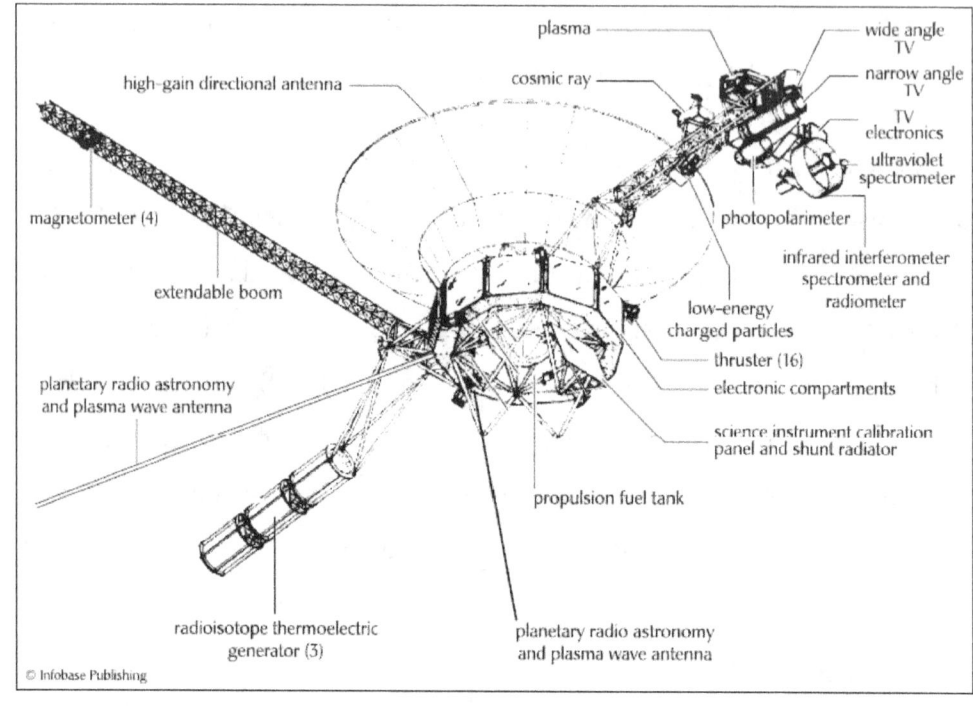

Figure 3.5: Scientific instruments aboard Voyager 2

Once every 176 years, the giant outer planets—Jupiter, Saturn, Uranus, and Neptune—align themselves in such an orbital pattern that a spacecraft launched from Earth to Jupiter at just the right time might be able to visit the other three planets on the same mission, by using a gravity-assist of these planets. *Voyager 2* took advantage of this celestial-alignment opportunity in 1977 and made planetary exploration history.

Launched in 1977, the twin *Voyager 1* and *Voyager 2* spacecraft flew by the planets Jupiter (1979) and Saturn (1980–81). *Voyager 2* then went on to have an encounter with Uranus (1986) and with Neptune (1989). Both *Voyager 1* and *Voyager 2* are now traveling on different trajectories into interstellar space. The Voyager Interstellar Mission (VIM) should continue well into the next decade.

Each Voyager spacecraft had a mass of 825 kg and carried a complement of scientific instruments to investigate the outer planets and their moons and intriguing ring systems. These instruments were powered by electric power by a long-lived, nuclear power system called a radioisotope thermoelectric generator (RTG)

Taking advantage of the 1977 launch window, the *Voyager 2* spacecraft was launched on August 20, 1977, on board a Titan-Centaur rocket. *Voyager 1* was launched on September 5, 1977. This spacecraft followed the same trajectory as its *Voyager 2* twin and overtook Voyager 2 just after entering the asteroid belt in mid-December 1977. *Voyager 1* made its closest approach to Jupiter on March 5, 1979, and then used Jupiter's gravity to swing itself to Saturn. On November 12, 1980, *Voyager 1* successfully encountered the Saturn system and then was flung up out of the ecliptic plane on an interstellar trajectory.

The *Voyager 2* spacecraft encountered the Jupiter system on July 9, 1979, and then used the gravity-assist technique to follow *Voyager 1* to

Saturn. On August 25, 1981, *Voyager 2* encountered Saturn. Voyager 2, using gravity assist from Saturn went on to successfully encounter both Uranus (January 24, 1986) and Neptune (August 25, 1989).

Chapter 4: Neptune and General Spacecraft Configuration

The purpose of this chapter is to discuss general details of the mission to Neptune. The following information provides leads to setting up the mission requirements. It discusses the possibilities of launching a mission to Neptune and the problems that will arise during the design of the mission and thus, shape the spacecraft according to the needs. The chapter mainly focuses on details about planet Neptune, the onboard instruments that will be required to analyse the planet, the power subsystem that will power the spacecraft and the propulsion subsystem that can propel the spacecraft to such a large distance from Earth in a significantly low period of time.

4.1. Neptune

Neptune is the eighth and farthest planet from the Sun in our Solar System. It is the fourth-largest planet in diameter and the third-largest planet by mass. The mass of Neptune is approximately 17 times the mass of Earth. The average distance of Neptune's orbit from the Sun is 30.1 AU, approximately 30 times the Earth-Sun distance. Due to the large distance from the Earth, it has remained much unexplored and obscure. Neptune has been visited by only one spacecraft, Voyager 2, which flew by the planet on August 25, 1989. Neptune is similar in composition to Uranus, and both have compositions which differ from those of

53

the larger gas giants Jupiter and Saturn. The interior of Neptune, like that of Uranus, is primarily composed of ices and rock. Traces of methane in the outermost regions in part account for the planet's blue appearance.

4.1.1. Surface and Weather Conditions on Neptune:

Neptune is a gas giant planet. It does not have a solid surface. The blue-green atmosphere observed in photographs and images of Neptune is really the top of the clouds on Neptune. As we move inside the atmosphere of Neptune, we will find an interior with increasing temperatures and pressures right down to the rocky core at the centre. Because of its great distance from the Sun, Neptune's outer atmosphere is one of the coldest places in the Solar System, with temperatures at its cloud tops approaching −218 °C (55 K).

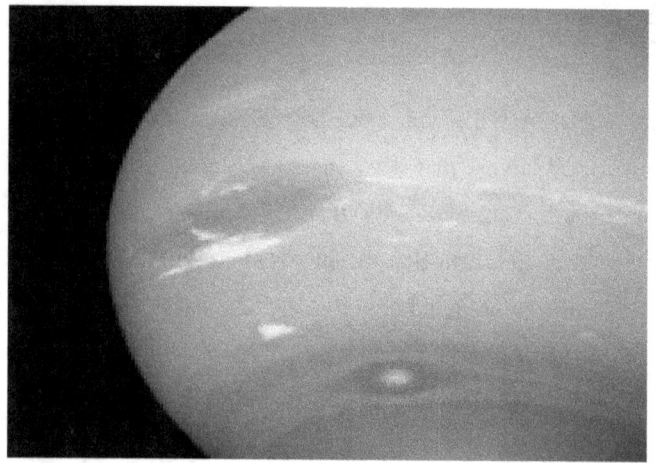

Figure 4.1: Atmosphere of Neptune

Temperatures at the planet's centre, however, are approximately 5,400 K (5,000 °C). The weather on Neptune is some of the most violent weather in the Solar System. The winds on Neptune travel at 2,100 km/hour. It is possible that the cold temperatures might decrease the friction in the system, so that winds can get going fast on Neptune. The more active weather on Neptune might be due to its higher internal heat. Neptune radiates 2.61 times as much energy as it receives from the Sun. This is enough heat to help drive the fastest winds in the Solar System. The following table provides a collection of general data for Neptune.

4.1.2. Exploration of Neptune:
The exploration of Neptune has only begun with one explorer, Voyager 2, which visited on August 25, 1989. The possibility of a Neptune Orbiter was discussed, yet other than that, no other missions have been given serious thought. As Neptune is a gas giant and has no solid surface, a surface mission such as a lander or rover is impossible.

Average Distance from Sun:	4.498 x 109km
Eccentricity of Orbit:	0.010
Average Orbital Speed:	5.5 km/s
Orbital Period:	164.86 years
Rotational Period:	16.11 hours
Inclination of Equator to Orbit	29.56°
Diameter:	49,528 km
Mass:	1.024 x 1026 kg
Average Density:	1638 kg/m3
Escape Speed:	23.5 km/s
Average Cloud-Top Temperature:	-218° C
Atmospheric Composition:	79% hydrogen
	18% helium
	3% methane

Other than Voyager 2, no other mission to Neptune has been under serious consideration. This is because of the great distance at which Neptune is from the Earth. Propelling a spacecraft to Neptune with the use of chemical or nuclear propulsion is a possibility. But it would require a huge propellant payload or a really big mechanism and even with a flyby from Jupiter, the time involved in the flight would be more than 20 years. In Voyager 2's last planetary encounter, the spacecraft swooped only 3,000 miles above Neptune's north pole, the closest approach it made to any celestial body since it left Earth. Voyager 2 utilized a radioisotope thermoelectric generator (RTG) for propulsion. It took Voyager 2 only 12 years to reach Neptune. As compared to the capability of chemical propulsion this seems strange. The reduced time of flight of Voyager 2 spacecraft was a result of reoccurring flyby's due to the unique celestial alignment of Saturn, Jupiter, Uranus and Neptune.

Voyager 2 studied Neptune's atmosphere, Neptune's rings and its magnetosphere. It also studied Neptune's moons. Several discoveries were made including the discovery of the Great Dark Spot and Triton's geysers. Voyager 2 revealed that Neptune's atmosphere was very dynamic, even though it receives only 3% of the sunlight Jupiter receives. Voyager 2 discovered an anticyclone called the Great Dark Spot, similar to Jupiter's Great Red Spot and Little Red Spot. However, images taken by the Hubble Space Telescope revealed that the Great Dark Spot had disappeared. Also seen in Neptune's atmosphere at that time was an almond-shaped spot designated D2, and a bright, quickly moving cloud high above the cloud decks dubbed "Scooter".

Voyager 2 spacecraft:

Voyager 2 found four rings and evidence for ring arcs, or incomplete rings above Neptune. Voyager 2 also studied Neptune's magnetosphere. The planetary radio astronomy instrument found that Neptune's day lasts sixteen hours, seven minutes. Voyager 2 also discovered auroras, like on Earth, but much more complex. Voyager 2 discovered six moons orbiting Neptune, but only three were photographed in detail: Proteus, Nereid, and Triton. Proteus turned out to be an ellipsoid, as large as an ellipsoid could become without rounding in a sphere. Proteus is very dark in color, almost like soot.

Nereid, though discovered in 1949, still has very little known about it. Triton was flown by at about 25,000 miles away, and became the last solid world Voyager 2 would explore within the Solar System. Triton was revealed to have remarkable active geysers and polar caps. A very thin atmosphere was found, as well as thin clouds. Since Voyager 2 was successful exploring Neptune, our spacecraft would also contain a similar composition of subsystems and scientific instruments. Although, nuclear propulsion is replaced by ion propulsion and the power source changed to a nuclear source.

4.2. Spacecraft Instruments

The spacecraft utilizes a three-axis stabilized system that uses celestial or gyro referenced attitude control to maintain pointing of the high-gain antennas toward Earth. The prime mission science payload consisted of 10 instruments which conduct 11 investigations, including radioscience. Only five investigator teams are still supported. With the exception of the Voyager 1's PLS instrument, all other are working well and are capable of continuing operations in the expected environment. In addition, data is collected from the Planetary Radio Astronomy (PRA) instrument and Voyager 1's Ultraviolet Spectrometer (UVS). The Flight Data Subsystem (FDS) and a single 8-track Digital tape recorder (DTR) provide the data handling functions. The FDS configures each instrument and controls instrument operations. It also collects engineering and science data and formats the data for transmission. The DTR is used to record high-rate PWS data. Data is played back every six months.

The command computer subsystem (CCS) provides sequencing and control functions The CCS contains fixed routines such as command decoding and fault detection and corrective routines, antenna pointing information, and spacecraft sequencing information.

59

The Attitude and Articulation Control Subsystem (AACS) controls spacecraft orientation, maintains the pointing of the high gain antenna towards Earth, controls attitude maneuvers, and positions the scan platform. Uplink communications is via S-band (16-bits/sec command rate) while an X-band transmitter provides downlink telemetry at 160 bits/sec normally and 1.4 kbps for playback of high-rate plasma wave data. All data are transmitted from and received at the spacecraft via the 3.7 meter high-gain antenna (HGA).

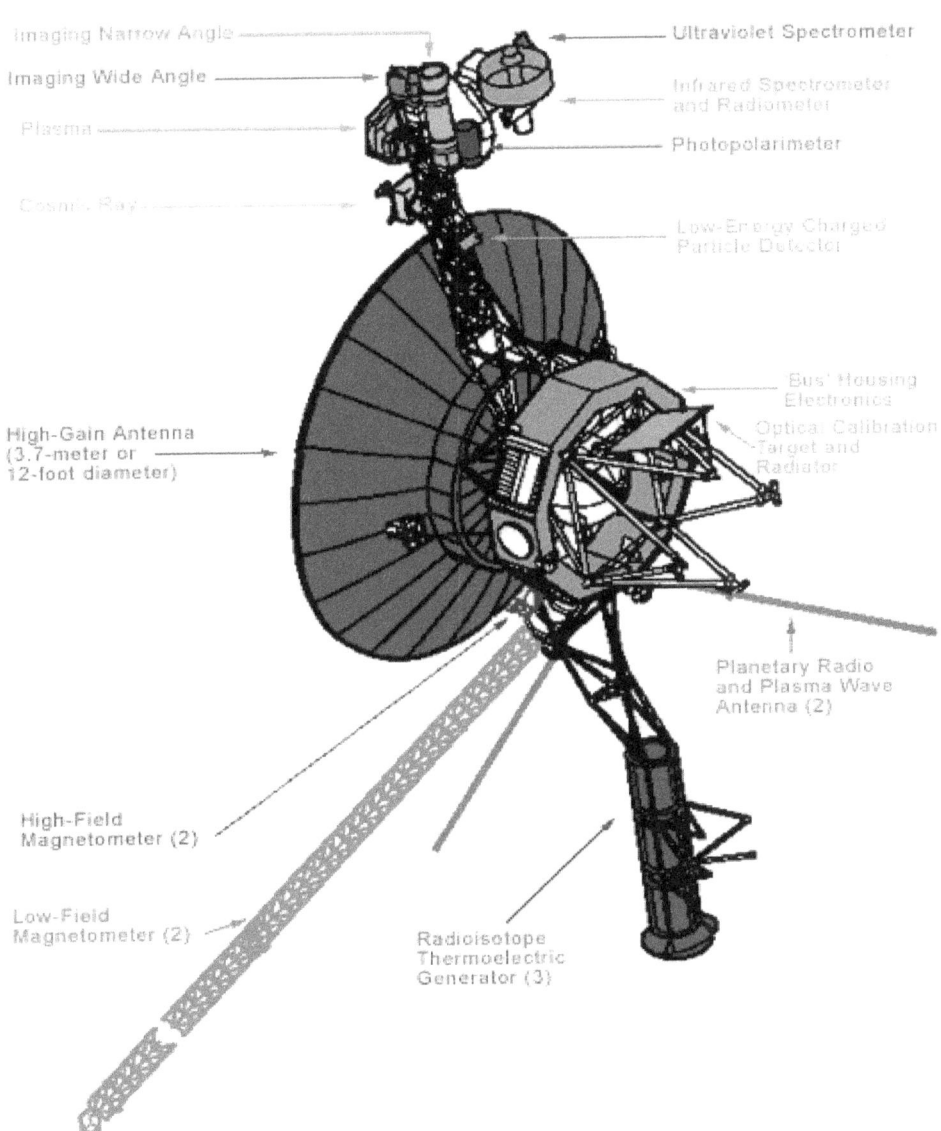

Imaging Narrow Angle
Imaging Wide Angle
Plasma
Cosmic Ray

Ultraviolet Spectrometer
Infrared Spectrometer and Radiometer
Photopolarimeter
Low-Energy Charged Particle Detector

High-Gain Antenna (3.7-meter or 12-foot diameter)

Bus' Housing Electronics
Optical Calibration Target and Radiator

Planetary Radio and Plasma Wave Antenna (2)

High-Field Magnetometer (2)

Low-Field Magnetometer (2)

Radioisotope Thermoelectric Generator (3)

Figure 4.2: Various spacecraft instruments

A range of scientific instruments were used in Voyager 2. A few of these instruments are tabulated below:

Table 2 : List of Scientific Instruments on-board Voyager 2

Investigator Teams	Instrument Measurements
Plasma Science (PLS)	Properties and radial evolution of the solar wind (ions 10 eV - 6 keV, electrons 4 eV-6 keV)
Low-Energy Charged Particles(LECP)	Energy spectrum of low-energy particles (electrons 10-10,000 keV, ions 10-150,000 keV/n)
Cosmic Ray Sub-system (CRS)	Energy spectrum of high- and low-energy electrons (3-110 MeV) and cosmic ray nuclei (1-500 MeV/n)
Magnetometer (MAG)	High (50,000 - 200,000 nT) and low (8-50,000 nT) magnetic field intensity
Plasma Wave Subsystem (PWS)	Electrical field components of plasma waves in frequency range of 10 Hz to 56 kHz

It is due to the large distance from Earth that Neptune is unexplored and obscure. Due to the enormous distance from Sun the intensity of Energy received by Neptune from the Sun in form of light is scarce. Hence there are many questions when it comes to launching a feasible mission to Neptune. However, the two major problems that hinder the launch are the technologies that will be used to propel and power the spacecraft.

4.3. Propulsion Subsystems

Spacecraft are provided with sets of propulsive devices so they can maintain three-axis stability, control spin, execute maneuvers, and make minor adjustments in trajectory. The more powerful devices are the engines that provide a force of several hundred Newton. These may be used to provide the large torques necessary to maintain stability during a solid rocket motor burn, or they may be used for orbit insertion. The spacecraft is also provided with smaller sets of rockets, generating between less than 1 N and 10 N, which are typically used to provide the delta-V for interplanetary trajectory correction maneuvers, orbit trim maneuvres, reaction wheel desaturation maneuvers, or routine three-axis stabilization or spin control. Other components of propulsion subsystems include propellant tanks, plumbing systems with electrically or pyrotechnically operated valves, and helium tanks to supply pressurization for the propellant. Some propulsion subsystems, such as Galileo's, use hypergolic propellants--two compounds stored separately which ignite spontaneously upon being mixed in the engines or thrusters. Other spacecraft use hydrazine, which decomposes explosively when brought into contact with an electrically heated metallic catalyst. Cassini, whose propulsion system is illustrated below, uses both hypergolics for its main engines and hydrazine monopropellant for its thrusters.

Not many types propulsion subsystems have been applied for the use of space propulsion. Up to now, the two majorly successful propulsion systems have been Solar and nuclear propulsion. For mission within the orbit of Mars, the use of a photo-voltaic solar panel is efficient as the intensity of energy from the sun is considerable. Although, beyond Mars, the use of solar power becomes unfeasible and use of a radioisotope thermoelectric generator becomes feasible. Even though the specific impulse achieved through nuclear propulsion is significantly good, it requires a really long time to cover much greater distances, such as that to Neptune. Ion propulsion best suits missions to the edge of the solar system and beyond due to a very high impulse of nearly 10000 seconds.

The Deep Space 1 spacecraft was a pioneer in the use of ion-electric propulsion in interplanetary space. With their high specific impulse (due to high nozzle exit velocities), ion engines can permit spacecraft to achieve the high velocities required for interplanetary or interstellar flight.

DS-1 ION ENGINE:

The principle behind an ion engine is the ionization of a gas such as xenon to make it responsive to electric and magnetic fields. Then the ions are then accelerated to extremely high velocity using electric fields and ejected from the engine. Electrical power comes from arrays of photovoltaic cells converting sunlight, so the technology is also called solar-electric propulsion. The ejecting of mass at extremely high speed provides the force against which the reaction is spacecraft acceleration in the opposite direction. The much higher exhaust speed of the ions as compared to chemical or nuclear rocket exhaust is the main factor in the engine's higher performance. The ion engine also emits electron so as to avoid building a negative electric charge on the spacecraft and causing the positively charged ion clouds to follow it.

4.3.1. Ion Propulsion

The efficient use of fuel and electrical power by an ion propulsion system, enables modern spacecrafts to travel farther, faster, and cheaper than any other propulsion technology currently available. Ion thrusters are currently used for stationkeeping on communication satellites and for main propulsion on deep space probes. Ion thrusters expel ions to create thrust and can provide higher spacecraft top speeds than any other rocket currently available.

An ion is simply an atom or molecule that is electrically charged. Ionization is the process of electrically charging an atom or molecule by adding or removing electrons. Ions can be positive or negative. Plasma is an electrically neutral gas in which all positive and negative charges--from neutral atoms, negatively charged electrons, and positively charged ions--add up to zero. Plasma is designated as the fourth state of matter (the others are solid, liquid, and gas) and it exists everywhere in nature. It has some of the properties of a gas but is affected by electric and magnetic fields and is a good conductor of electricity. Plasma is the building block for all types of electric propulsion. Electric and/or magnetic fields are used to push on the electrically charged ions and electrons to provide thrust. Examples of plasmas seen every day are lightning and fluorescent light bulbs. The conventional method deployed for ionizing the propellant atoms in an ion thruster is called electron bombardment. The principle behind electron bombardment is, when a high-energy electron (negative charge) collides with a propellant atom (neutral charge), a second electron is released, yielding two negative electrons and one positive ion.

An alternative method of ionization called electron cyclotron resonance (ECR) is also being researched at NASA. The use of high-frequency radiation (usually microwaves) coupled with a high magnetic field enables to heat the electrons in the propellant atoms, causing them to break free of the propellant atoms, creating plasma. Ions can then be extracted from this plasma.

4.3.2. Ion Thruster Operation

Modern ion thrusters use inert gases for propellant. The majority of thrusters use xenon, which is chemically inert, colorless, odorless, and tasteless. The propellant is injected from the downstream end of the thruster and flows toward the upstream end. This injection method is preferred because it increases the time that the propellant remains in the chamber.

In a conventional ion thruster, electrons are generated by a hollow cathode, called the discharge cathode, located at the center of the engine on the upstream end. The electrons flow out of the discharge cathode and are attracted to the discharge chamber walls, which are charged to a high positive potential by the thruster's power supply.The propellant is ionized by the electrons from the discharge cathode by means of electron bombardment. High-strength magnets redirect the electrons approaching the walls, where they are placed, into the discharge chamber by the magnetic fields. By maximizing the length of time that electrons and propellant atoms remain in the discharge chamber, the chance of ionization is maximized, which makes the ionization process as efficient as possible.

In an ion thruster, ions are accelerated by electrostatic forces. The electric fields used for acceleration are generated by electrodes positioned downstream end of the thruster. Each set of electrodes, called ion optics or grids, contains thousands of coaxial apertures. Each set of apertures acts as a lens that electrically focuses ions through the optics. Ion thrusters usually use a two-electrode system, where the upstream electrode (called the screen grid) is charged highly positive, and the downstream electrode (called the accelerator grid) is charged highly negative. The ions, generated in a region of high positive, are attracted toward the accelerator grid and are focused out of the discharge chamber through the apertures, creating thousands of ion jets. The stream of all the ion jets together is called the ion beam. The thrust force is the force that exists between the upstream ions and the accelerator grid. The exhaust velocity of the ions in the beam is based on the voltage applied to the optics. While a chemical rocket's top speed is limited by the thermal capability of the rocket nozzle, the ion thruster's top speed is limited by the voltage that is applied to the ion optics (which is theoretically unlimited).

Due to the large amount of positive ions expelled by an ion thruster, an equal amount of negative charge must be expelled to keep the total charge of the exhaust beam neutral. This is done by a second hollow cathode called the neutralizer is located on the downstream perimeter of the thruster and expels the needed electrons.

4.3.3. The Electric Propulsion System

The ion propulsion system (IPS) consists of five main parts:

1. The Power Source:

The IPS power source can be any source of electrical power. But due to the limiting conditions of operation and quantity, solar and nuclear are the primary options. A solar electric propulsion system (SEP) utilizes sunlight and solar cells for power generation. A nuclear electric propulsion system (NEP) uses an electric generator coupled to a nuclear heat source.

2. Power Processing unit (PPU),

The PPU converts the electrical power generated by the power source into the power required for each component of the ion thruster. It generates the voltages required by the ion optics and discharge chamber. It also provides voltage for the high currents required for the hollow cathodes.

3. The Control Computer

The control computer controls and monitors system performance.

4. Propellant Management System (PMS)

The PMS controls the propellant flow from the propellant tank to the thruster and hollow cathodes. Modern PMS units have evolved to a level of sophisticated design that no longer requires moving parts.

5. The Ion Thruster

The ion thruster then processes the propellant and power to perform work. Modern ion thrusters are capable of propelling a spacecraft up to 90,000 meters per second (over 200,000 miles per hour (mph). The space shuttle is capable of an incredible speed of around 18,000 mph. The tradeoff for this high top speed is low thrust (or low acceleration). Modern ion thrusters can deliver up to 0.5 Newtons (0.1 pounds) of thrust. To compensate for low thrust, the ion thruster must be operated for a long time for the spacecraft to reach its top speed. But the advantages of Ion thrusters compensate this lack of thrust. Ion thrusters use inert gas for propellant, eliminating the risk of explosions associated with chemical propulsion. The usual propellant is xenon, but other gases such as krypton and argon may be used.

Past:

The NASA Glenn Research Center has been the lead for electric propulsion since work on ion propulsion began there in the 1950s. Space Electric Rocket Test 1 (SERT 1) was the first operational test of an ion propulsion system in space. It flew on July 20, 1964, and successfully completed its goal of 31 minutes of operation before its return to Earth. Many successful tests of ion propulsion followed. The Ion Auxiliary Propulsion System (IAPS) project from 1974 to 1983 developed an 8-centimeter mercury IPS for satellite station keeping. A 30-centimeter IPS that was used as the main propulsion on the Deep Space 1 (DS1) spacecraft from 1998 to 2001 was developed by the NASA Solar Technology Application Readiness (NSTAR). DS1 was the first use of electric propulsion for spacecraft main propulsion. The NSTAR thruster on DS1 propelled the spacecraft 263,179,600 kilometers (163,532,236 miles) at speeds up to 4,500 meters per second (10,066 mph). Over the entire mission, the NSTAR thruster demonstrated 200 starts and 16,246 hours of operation.

Present:

The NASA Evolutionary Xenon Thruster (NEXT) project is developing a high-power SEP IPS that will reduce mission cost and trip time. Mars and Saturn can be targeted by NEXT. It is capable of performing a wide variety of missions to targets. In 2003, a single NEXT thruster demonstrated over 2000 hours of operation at 7 kilowatts. The NEXT single string integration test, also completed in 2003, demonstrated operation of the NEXT PPU, PMS, and thruster as a complete IPS. The High Power Electric Propulsion (HiPEP) project is developing a high-power NEP IPS for the Jupiter Icy Moons Orbiter (JIMO) spacecraft. The HiPEP thruster, which is under development at Glenn is unique. It has the ability to operate at high power levels in both a conventional hollow cathode configuration and a microwave ECR configuration. The HiPEP ion thruster is currently the most powerful inert gas ion thruster ever built. Early tests in 2004 demonstrated power levels of 40 kilowatts and exhaust velocities in excess of 90,000 meters per second (over 200,000 mph).

Future:

More and more companies are beginning to use satellites with electric propulsion to extend the operational life of satellites and reduce launch and operation costs. Primary application of ion propulsion will be for providing propulsion on long missions that are difficult or impossible to perform using other types of propulsion. The Dawn spacecraft, scheduled for launch in May 2006 will study Ceres and Vesta, two protoplanets located in the asteroid belt that exists between Mars and Jupiter. It uses three NSTAR ion thrusters as main propulsion. By studying these protoplanets, which were among the first bodies formed in our solar system, researchers hope to gain valuable information about the solar system's early development.

The JIMO spacecraft using an array of high-power ion thrusters as main propulsion, will perform an extensive exploration of Jupiter's icy moons Callisto, Ganymede, and Europa. The spacecraft will investigate each moon's composition, history, and potential for sustaining life. Research in the area of ion propulsion continues to push the envelope of propulsion technology. Advancements are being made that allow the thrusters to operate at higher power levels, higher speeds, and for longer durations. PPU and PMS technologies are being developed that will allow the development of lighter and more compact systems while increasing reliability. Higher power thrusters will be developed that provide greater speed and more thrust, as new power sources become available. Supporting technologies such as carbon-based ion optics and ECR discharges may greatly increase ion thruster operational life. It will allow longer duration missions or high-power IPS operation. These technologies will allow humankind to explore the farthest reaches of our solar system

TRAJECTORY ANALYSIS

The purpose of this section is to present the analysis of the trajectory of the spacecraft from planet Earth to planet Neptune. The problem is presented in the form of an interplanetary non-Hohmann transfer of a scientific space probe from Earth to Neptune. Based on the propulsion system used for the spacecraft, the time flight of the spacecraft from Earth to Neptune is estimated to be ten years.

The objective is to calculate the month and year of launch for the total change in velocity required, i.e., at Earth to put the spacecraft into a heliocentric trajectory to Neptune and at Neptune to set it into parking orbit around Neptune, is the least. The program generated calculates the corresponding month and year of launch, from January, 2011 to December, 2099, for which the total delta-V is minimum. It also returns the orbital elements of the most optimal interplanetary non-Hohmann trajectory that the spacecraft follows.

The assumptions made for the development of the model are as follows:

1. The spacecraft is assumed to be launched from a parking orbit above the Earth's surface.

2. The model assumes method of patched conics. Thus the calculation can be made considering a heliocentric path from the position vector of Earth at the time of departure to the position vector of Neptune at the time of arrival.

3. It is assumed that the spacecraft during its flight trajectory from Earth to Neptune does not come in the sphere of influence of any other planet.

The steps involved in the solution to the problem along with theoretical background are presented in the following sections. The corresponding results from the m-files generated for the solution are presented along with the steps.

Chapter 5: STATE VECTOR

The first step involved in the solution to the problem is the calculation of the state vector of the two planets at variable instants of time, i.e., every month from 2011 to 2099. For calculation of state vector of a celestial body (Earth or Neptune) under the gravitational influence around another body(The Sun) requires the knowledge of the "two body problem" which describes the motion of two bodies under gravitational influence.

5.1. The Two Body Problem

Since the motion of the planets and the spacecraft are trajectories around the Sun or Heliocentric paths, thus for the model developed, the position reference system is taken to be an inertial frame of reference with the Sun as the origin.

5.1.1. Equations of motion in Inertial frame of reference

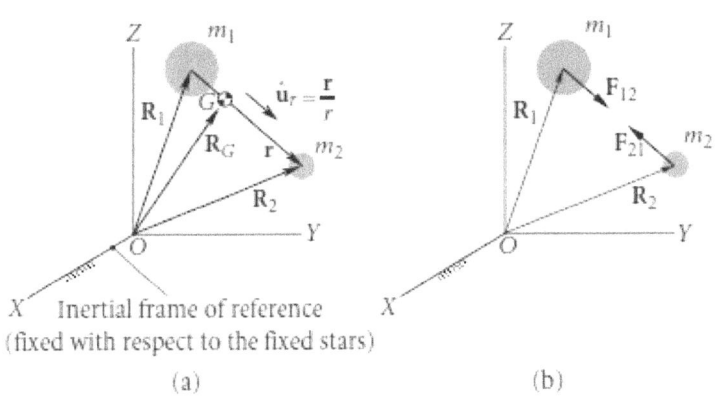

Figure 5.1: Force of gravitation

Figure 6.1 shows two point masses acted upon only by the mutual force of gravity between them. The positions of their centers of mass are shown relative to an inertial frame of reference XYZ. The origin O of the frame may move with constant velocity (relative to the fixed stars), but the axes do not rotate. Each of the two bodies **is** acted upon by the gravitational attraction of the other. $\mathbf{F}_{12}$ is the force exerted on m_1 by m_2, and $\mathbf{F}_{21}$ is the force exerted on m_2 by m_1.

Let $\mathbf{r}$ be the position vector of m_2 relative to m_1. Then

$$r = R_2 - R_1 \qquad (5.1)$$

77

The body m_1 is acted upon only by the force of gravitational attraction towards m_2. The force of gravitational attraction, F_g , which acts along the line joining the centers of mass of m_1 and m_2. The force exerted on m_2 by m_1 is

$$F_{21} = -\frac{Gm_1m_2}{r^2}\hat{u}_r$$

(5.2)

Where $\hat{u}_r$ is the unit vector pointing from m_1 towards m2 and $-\hat{u}_r$ accounts for the fact that the force vector $\mathbf{F}_{21}$ is directed from m_2 towards m_1. Newton's second law of motion as applied to body m_2 is

$$F_{21} = m_2\ddot{R}_2$$

(5.3)

Where $\ddot{R}_2$ is the absolute acceleration of m_2. Thus

$$-\frac{Gm_1m_2}{r^2}\hat{u}_r = m_2\ddot{R}_2$$

(5.4)

By Newton's third law (the action–reaction principle), $\mathbf{F}_{12} = -\mathbf{F}_{21}$, so that for m_1 we have

$$\frac{Gm_1m_2}{r^2}\hat{u}_r = m_2\ddot{R}_1$$

(5.5)

5.1.2. Equations of Relative motion

The position vector of $\mathbf{R_2}$ relative to $\mathbf{R_1}$ is given by $\mathbf{r}$. It can be obtained by multiply Equation 2.7 by m_1 and Equation 2.8 by m_2 and subtracting the second of these two equations from the first. It yields:

$$\ddot{r} = -\frac{\mu}{r^3}r \qquad (5.6)$$

The gravitational parameter μ can be defined as

$$\mu = G(m_1 + m_2) \qquad (5.7)$$

The units of μ are km^3s^{-2}. Equation 2.13 is a second order differential equation that governs the motion of m_2 relative to m_1. The specific angular momentum of body m_2 relative to m_1 is given by

$$h = r \times \dot{r} \qquad (5.8)$$

The specific relative angular momentum of a body in orbit around another body is constant. Therefore,

$$\frac{dh}{dt} = 0 \qquad (5.9)$$

The relationship between $\mathbf{r}$ and $\mathbf{h}$ can be represented in the following form (chapter 2, reference 3):

79

$$\frac{r}{r} + r.e = \frac{\ddot{r}.xh}{\mu}$$

(5.10)

Where, dimensionless vector **e** is called the eccentricity vector. The line defined by the vector **e** is commonly called the apse line. In order to obtain a scalar equation, we take the dot product of both sides of Equation 2.30 with **r**. Utilizing vector identities, the vector equation can be used to obtain the scalar equation for 'r'.

$$r = \frac{h^2}{\mu} \frac{1}{1 + e cos\Theta}$$

(5.11)

This is the orbit equation, and it defines the path of the body m_2 around m_1, relative to m_1. μ, h, and e are constants for a particular trajectory. Where e is the magnitude of the eccentricity vector **e** and θ or true anomaly is the angle between the fixed vector **e** and the variable position vector **r**.

5.2. Types of Orbits

5.2.1. Elliptical Trajectories (0<e<1)
If 0<e <1, then the denominator of Equation 2.35 varies with the true anomaly θ, but it remains positive, never becoming zero. Therefore, the

relative position vector remains bounded, having its smallest magnitude at point P (called periapsis), r_p and the largest magnitude at point A (called appoapses), r_a.

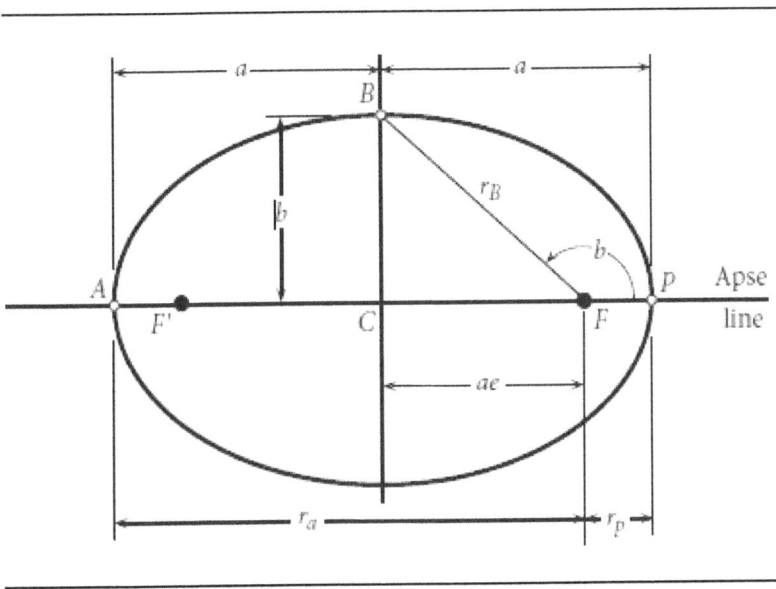

Figure 5.2: Elliptical trajectory

F and F' are the two foci of the ellipse. During elliptical movement of a body (m_2) under the gravitational influence of another(m_1), the real focus F is occupied by m1 and the imaginary focus F' is empty. The semi- major axis of the ellipse is 'a'. Energy of the elliptical trajectory is given by:

$$\frac{v^2}{2} - \frac{\mu}{r} = -\frac{\mu}{2a}$$

(5.12)

The mean anomaly is the azimuth position (in radians) of a fictitious body moving around the ellipse at the constant angular speed n.

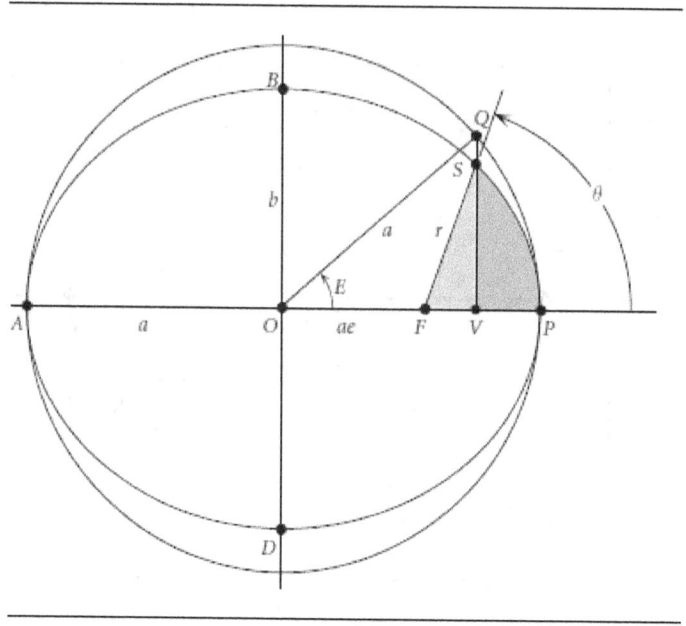

Figure 5.3: Eccentricity anomaly

E is an auxiliary angle called the eccentric anomaly, shown in Figure 3.3. This is done by circumscribing the ellipse with a concentric auxiliary circle having a radius equal to the semimajor axis 'a' of the ellipse. Through geometry, the auxiliary angle E in terms of theta is given by:

$$cosE = \frac{e + cos\theta}{1 + ecos\theta}$$ (5.13)

Using trigonometric identity, the mean anomaly M_e can be written in terms of eccentricity anomaly E and true anomaly θ as:

$$M_e = E - esinE$$ (5.14)

This is Kepler's equation. The solution to Kepler's equation for the value of eccentricity anomaly E given the values of mean anomaly M and eccentricity e is done iteratively using Newton's method. The MATLAB algorithm for solution of Kepler's equation is given in the script "kepler_E.m".

Now we can define the major equations for an ellipse in a tabulated form. In a similar fashion, the corresponding equations for hyperbolic trajectories are tabulated in the respective sections.

In tabulated form:

Table 3: Equation for elliptical orbits

Equation	Ellipse
1. Orbit Equation	$r = \dfrac{h^2}{\mu} \dfrac{1}{1 + ecos\theta}$
2. Conic equation in cartesian coordinates	$\dfrac{x^2}{a^2} + \dfrac{y^2}{b^2} = 1$

83

3. Semimajor axis	$$a = \frac{h^2}{\mu} \frac{1}{1-e^2}$$
4. Semiminor axis	$$b = a\sqrt{1-e^2}$$
5. Energy equation	$$\frac{v^2}{2} - \frac{\mu}{r} = -\frac{\mu}{2a}$$
6. Kepler's equation	$$M_e = E - esinE$$

5.2.2. Parabolic Trajectories (e=1)

When the eccentricity equals 1, the orbit equation becomes:

$$r = \frac{h^2}{\mu} \frac{1}{1 + cos\theta}$$

(5.15)

As the true anomaly approaches 180^0, the denominator of equation 2.79 becomes zero. Thus, r approaches infinity. Parabolic trajectories are escape trajectories. If a satellite revolving around another body under the influence of gravitational force of attraction is to be sent out of the orbit of that body, a parabolic trajectory is used. When the spacecraft reaches infinity with respect to that body, its velocity is zero relative to the body. That is, the satellite in a parabolic trajectory possesses only that much amount of gravitational energy at a time instant, as it contains kinetic energy. The energy equation for a parabola becomes:

84

$$\frac{v^2}{2} - \frac{\mu}{r} = 0 \tag{5.16}$$

Thus, speed anywhere on the parabolic path is:

$$v = \sqrt{\frac{2\mu}{r}} \tag{5.17}$$

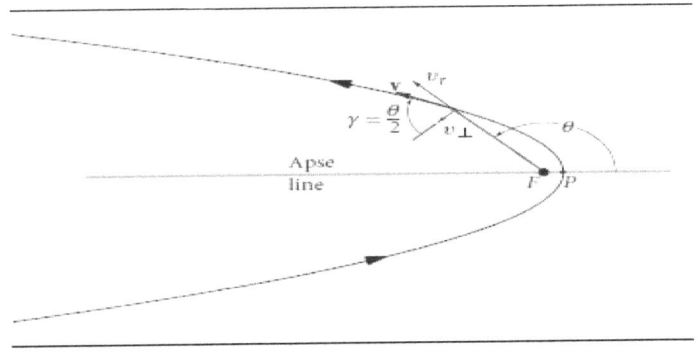

Figure 5.4: Parabolic Orbit

Thus at any distance r from m1, the escape velocity of m2 is given by

$$v_{esc} = \sqrt{\frac{2\mu}{r}} \tag{5.18}$$

Drawing a perpendicular to the apse line at F, the distance between the two points where the perpendicular cuts above and below the apse line is known as the latus-rectum or the parameter of the orbit.

5.2.3. Hyperbolic Trajectories (e>1)

For a hyperbolic trajectory, e > 1.The system consists of two symmetric curves. One of them is occupied by the orbiting body, the other one is its empty, mathematical image. Clearly, the denominator of Equation 2.86 goes to zero when $\cos \theta = -1/e$.

$$\theta_\infty = cos^{-1}(-\frac{1}{e})$$

(5.19)

θ_∞ is known as the true anomaly of the asymptote. When a body is sent into hyperbolic trajectory, it escapes the orbit around another body parallel to the asymptote.

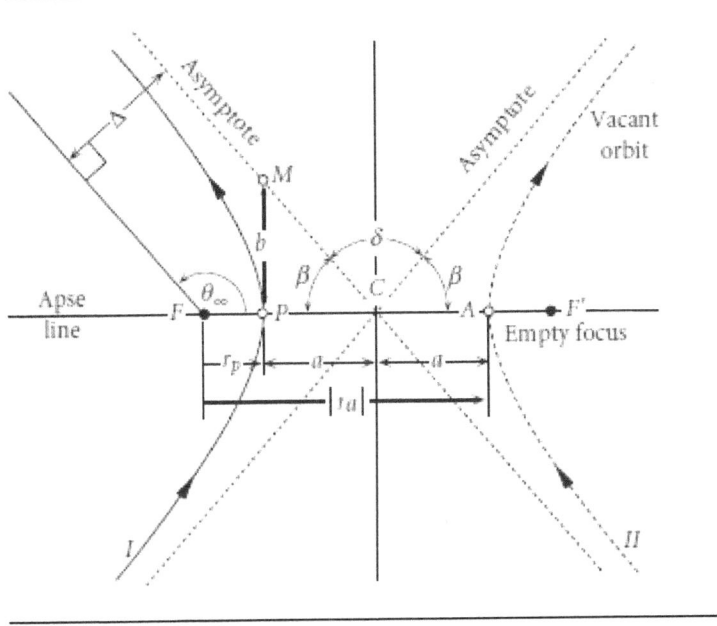

Figure 5.5: Hyperbolic Orbit

Hyperbolic excess velocity:

The specific energy of a hyperbolic orbit is positive and, similar to the elliptical orbit is given by the equation

$$\frac{v^2}{2} - \frac{\mu}{r} = \frac{\mu}{2a}$$

$$(5.20)$$

Let v_∞ denote the speed at which a body on a hyperbolic path arrives at infinity. Thus, according to equation 2.101,

$$v_\infty = \sqrt{\frac{\mu}{a}}$$

(5.21)

Since the energy is constant, Equation 2.101 may be written as

$$\frac{v^2}{2} - \frac{\mu}{r} = \frac{v_\infty^2}{2}$$

(5.22)

Since the escape velocity is given by

$$v_{esc} = \sqrt{\frac{2\mu}{r}}$$

(5.23),

we obtain for a hyperbolic trajectory

$$v^2 = v_{esc}^2 + v_\infty^2$$

(5.24)

This equation clearly shows that the hyperbolic excess speed v_∞ represents the excess kinetic energy over that which is required to simply escape from the center of attraction. The square of v_∞ is denoted C3, and is known as the characteristic energy, C3 is a measure of the energy required for an interplanetary mission and C3 is also a measure of the maximum energy a launch vehicle can impart to a spacecraft of a given mass.

Equations of a hyperbolic trajectory are tabulated as below:

Table 4: Hyperbolic trajectory equations

Equation	Hyperbola
1. Orbit Equation	$r = \dfrac{h^2}{\mu} \dfrac{1}{1 + e\cos\theta}$
2. Conic equation in cartesian coordinates	$\dfrac{x^2}{a^2} - \dfrac{y^2}{b^2} = 1$
3. Semimajor axis	$a = \dfrac{h^2}{\mu} \dfrac{1}{e^2 - 1}$
4. Semiminor axis	$b = a\sqrt{e^2 - 1}$
5. Energy equation	$\dfrac{v^2}{2} - \dfrac{\mu}{r} = \dfrac{\mu}{2a}$
6. Kepler's equation	$M_h = e\sinh F - F$

5.3. Orbits in Three Dimensions

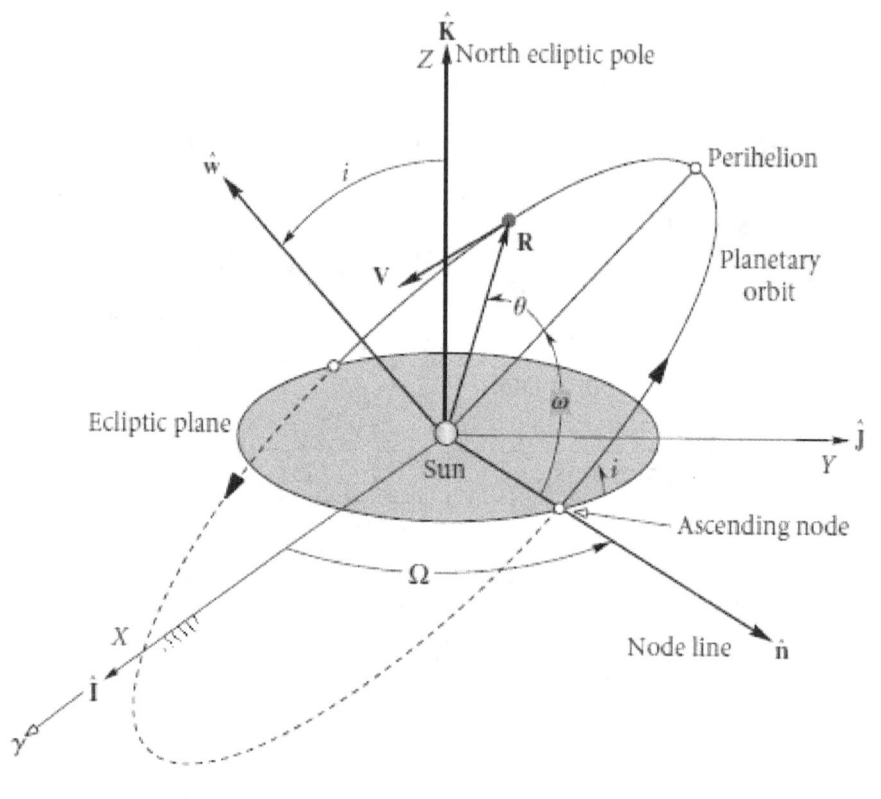

Figure 5.6: Planetary ephemeris

The state vector **R**, **V** of a planet is defined relative to the heliocentric ecliptic frame of reference illustrated in Figure 8.25. The sun is the centre of attraction. The ecliptic plane is the plane of the Earth revolution around the Sun. The Inertial X axis is defined by the vernal equinox. The coordinate system used to describe earth orbits in three dimensions is defined in terms of earth's equatorial plane, the ecliptic plane, and the earth's axis of rotation. The ecliptic is the plane of the earth's orbit around the sun, as illustrated in Figure 4.1. The earth's axis of rotation, which passes through the North and South Poles, is tilted away by an angle known as the obliquity of the ecliptic, approximately 23.4°. Therefore, the earth's equatorial plane and the ecliptic intersect along a line, which is known as the vernal equinox line. The position of the sun at the first day of spring in the northern hemisphere defines the location of a point in the sky called the vernal equinox, for which the symbol γ is used. Coordinates of latitude and longitude are used to locate points on the celestial sphere the same way as on the surface of the earth. The projection of the earth's equatorial plane outward onto the celestial sphere defines the celestial equator.

5.3.1. Orbital Elements
To define an orbit in the plane requires two parameters: eccentricity and angular momentum. Other parameters, such as the semimajor axis, the specific energy, and (for an ellipse) the period, are obtained from these two. To locate a point on the orbit requires a third parameter, the true anomaly.Describing the orientation of an orbit in three dimensions requires three additional parameters: i - the inclination to the ecliptic plane; Ω - the right ascension of the ascending node (relative to the J2000 vernal equinox); $\tilde{\omega}$ - the longitude of perihelion, is defined as $\tilde{\omega} = \omega + \Omega$, where ω is the argument of perihelion.

The vernal equinox γ, which lies on the celestial equator, is the origin for measurement of longitude, which in astronomical parlance is called right ascension. Right ascension (RA or Ω) is measured along the celestial equator in degrees east from the vernal equinox.

Looking at figure , the intersection of the orbital plane with the equatorial (XY) plane is called the node line. The point on the node line where the orbit passes above the equatorial plane from below it is called the ascending node and where the orbit dives below the equatorial plane, is the descending node. The angle between the positive X axis and the node line is the first Euler angle Ω, the right ascension of the ascending node. The dihedral angle between the orbital plane and the equatorial plane is the inclination i, measured according to the right-hand rule, that is, counterclockwise around the node line vector from the equator to the orbit. The inclination is a positive number between 0^0 and 180^0. The perigee of the orbit (perihelion) lies at the intersection of the eccentricity vector **e** with the orbital path. The third Euler angle ω, the argument of perigee, is the angle between the node line vector **N** and the eccentricity vector **e**, measured in the plane of the orbit. Thus, the six orbital elements are

h,- specific angular momentum

i- inclination

Ω- right ascension (RA) of the ascending node

92

e- eccentricity

ω- argument of perigee

θ-true anomaly

$\tilde{\omega}$ - the longitude of perihelion, is defined as $\tilde{\omega} = \omega + \Omega$

L, the mean longitude, is defined as $L = \tilde{\omega} + M$, where M is the mean anomaly described in section 1.1.2.

The J2000 orbital elements (*a, e, i, Ω, $\tilde{\omega}$ and L*) and their centennial rates $(\dot{a}, \dot{e}, \dot{i}, \dot{\Omega}, \dot{\tilde{\omega}} and \dot{L})$ for the nine planets of the solar system are tabulated below:

Table 5: J2000 orbital elements and centennial rates

	a, AU $\dot{a}$, AU/Cy	e $\dot{e}$,1/Cy	i, deg $\dot{i}$,"/Cy	Ω, deg $\dot{\Omega}$,"/Cy	$\tilde{\omega}$, deg $\dot{\tilde{\omega}}$,"/Cy	L, deg $\dot{L}$,"/Cy
Mercury	0.38709893 0.00000066	0.20563069 0.00002527	7.00487 −23.51	48.33167 −446.30	77.45645 573.57	252.25084 538 101 628.29
Venus	0.72333199 0.00000092	0.00677323 −0.00004938	3.39471 −2.86	76.68069 −996.89	131.53298 −108.80	181.97973 210 664 136.06
Earth	1.00000011 −0.00000005	0.01671022 −0.00003804	0.00005 −46.94	−11.26064 −18228.25	102.94719 1198.28	100.46435 129 597 740.63
Mars	1.52366231 −0.00007221	0.09341233 0.00011902	1.85061 −25.47	49.57854 −1020.19	336.04084 1560.78	355.45332 68 905 103.78
Jupiter	5.20336301 0.00060737	0.04839266 −0.00012880	1.30530 −4.15	100.55615 1217.17	14.75385 839.93	34.40438 10 925 078.35
Saturn	9.53707032 −0.00301530	0.05415060 −0.00036762	2.48446 6.11	113.71504 −1591.05	92.43194 −1948.89	49.94432 4 401 052.95
Uranus	19.19126393 0.00152025	0.04716771 −0.00019150	0.76986 −2.09	74.22988 −1681.4	170.96424 1312.56	313.23218 1 542 547.79
Neptune	30.06896348 −0.00125196	0.00858587 0.00002514	1.76917 −3.64	131.72169 −151.25	44.97135 −844.43	304.88003 786 449.21
Pluto	39.48168677 −0.00076912	0.24880766 0.00006465	17.14175 11.07	110.30347 −37.33	224.06676 −132.25	238.92881 522 747.90

MATLAB ALGORITHM #1: planetary_elements.m

This program extracts the J2000 orbital elements and their centennial rates from the table given above.

The J2000 elements are the orbit elements as their values exist at the start of the 20th century.

Variables:

Planet_id – Planet identifier

1 – Mercury

2 – Venus

3 – Earth

94

4 – Mars

5 – Jupiter

6 – Saturn

7 – Uranus

8 – Neptune

9 - Pluto

J2000_coe – Row matrix containing the orbital elements listed in the order as the above table

Rates – Row matrix containing the rates of orbital elements listed in the order as the above table

RESULTS:

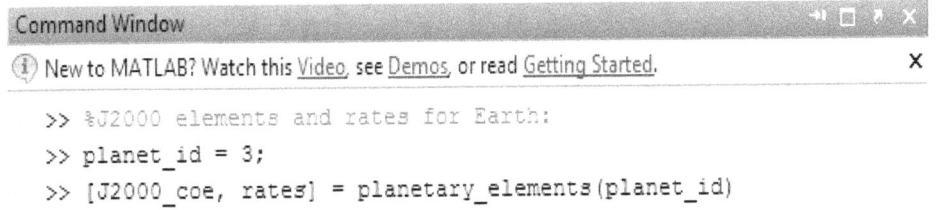

```
Command Window

New to MATLAB? Watch this Video, see Demos, or read Getting Started.

>> %J2000 elements and rates for Earth:
>> planet_id = 3;
>> [J2000_coe, rates] = planetary_elements(planet_id)
```

Variable Editor - J2000_coe1

Stack: Base   Select ...

J2000_coe1 <1x6 double>

	1	2	3	4	5	6
1	1.4960e+08	0.0167	5.0000e-05	-11.2606	102.9472	100.4644
2						
3						
4						
5						
6						

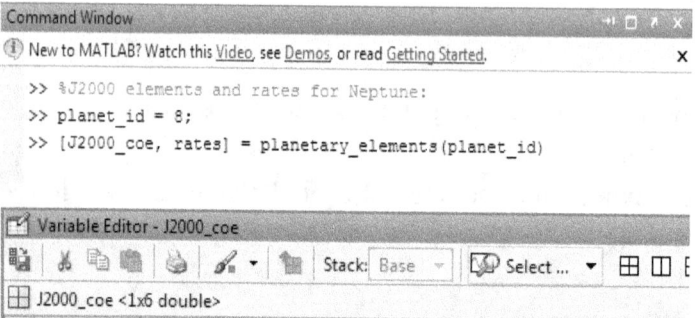

5.3.2 The Perifocal Frame

The perifocal frame is the 'natural frame' for an orbit. It is centered at the focus of the orbit. Its xy plane is the plane of the orbit, and its x axis is directed from the focus through periapse. The unit vector along the x axis is denoted by $\hat{p}$. The y axis, with unit vector $\hat{q}$, lies at 90^0 to the x axis. The z axis is normal to the plane of the orbit in the direction of the

angular momentum vector **h**. The z unit vector is $\hat{w}$. The perifocal frame
is illustrated in figure

In the perifocal frame, the position vector **r** is written as:

$$r = \bar{x}\hat{p} + \bar{y}\hat{q} \qquad\qquad (5.25)$$

Where

$$\bar{x} = r\cos\theta \qquad \bar{y} = r\sin\theta \qquad (5.26)$$

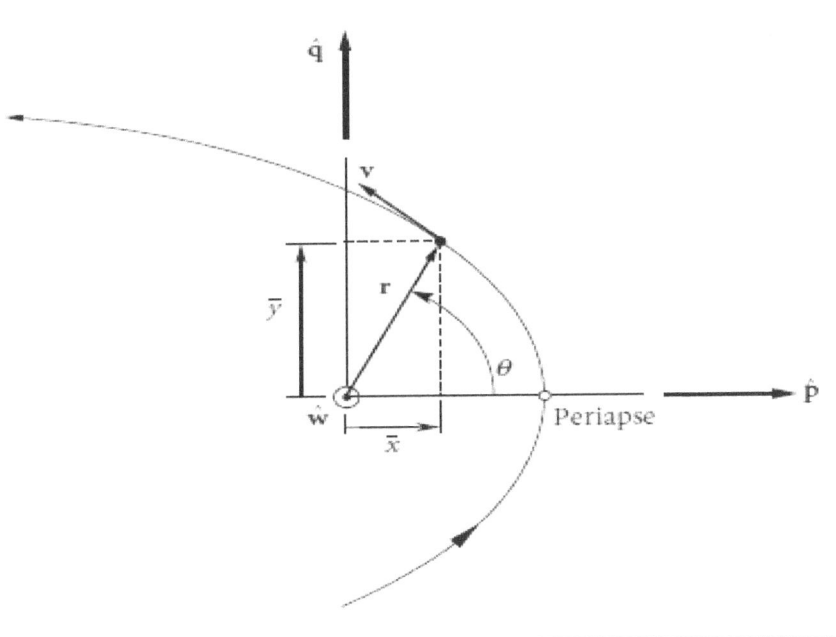

Figure 5.7: Perifocal frame

The velocity in perifocal frame is given by:

$$v = \dot{r} = \dot{x}\hat{p} + \dot{y}\hat{q} \qquad (5.27)$$

Where,

$$\dot{x} = -\frac{\mu}{h}\sin \qquad \dot{y} = -\frac{\mu}{h}(e + \cos\Theta) \qquad (5.28)$$

5.3.3. Transformation between Ecliptic and Perifocal frames

Figure 4.9 shows two cartesian coordinate systems: the system with axes xyz and orthogonal unit basis vectors $\hat{i}\hat{j}$ and $\hat{k}$; and the system with axes x'y'z' and orthogonal unit basis vectors $\hat{i}'$, $\hat{j}'$ and $\hat{k}'$.

Let [Q] represent the matrix of direction cosines of $\hat{i}\hat{j}$ and $\hat{k}$ relative to $\hat{i}'$, $\hat{j}'$ and $\hat{k}'$.

$$[Q] = \begin{bmatrix} Q_{11} & Q_{12} & Q_{13} \\ Q_{21} & Q_{22} & Q_{23} \\ Q_{31} & Q_{32} & Q_{33} \end{bmatrix} = \begin{bmatrix} \hat{i}.\hat{i}' & \hat{i}.\hat{j}' & \hat{i}.\hat{k}' \\ \hat{j}.\hat{i}' & \hat{j}.\hat{j}' & \hat{j}.\hat{k}' \\ \hat{k}.\hat{i}' & \hat{k}.\hat{j}' & \hat{k}.\hat{k}' \end{bmatrix} \qquad (5.28)$$

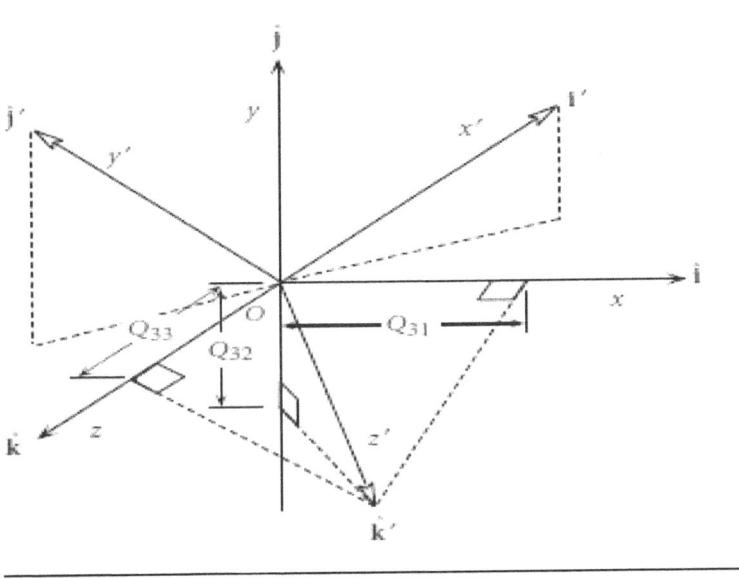

Figure 5.8: Transformation of frames of reference

The transpose of the matrix [**Q**], denoted by $[Q]^T$, is obtained by interchanging the rows and columns of [**Q**]. We see that the product

$$[Q]^T [\mathbf{Q}] = [\mathbf{1}] \qquad (4.26)$$

The transformation matrix [**Q**] is used to transform the position and velocity vector from one frame of reference to another through the following relationship:

$$\{r'\} = [Q]\{r\} \quad \{r\} = [Q]\{r'\} \qquad (5.29)$$

99

$$\{v'\} = [Q]\{v\} \quad \{v\} = [Q]\{v'\}$$ (5.30)

The transformation from the geocentric equatorial frame into the perifocal frame can be done by the sequence of three rotations illustrated in Figure 5.9.

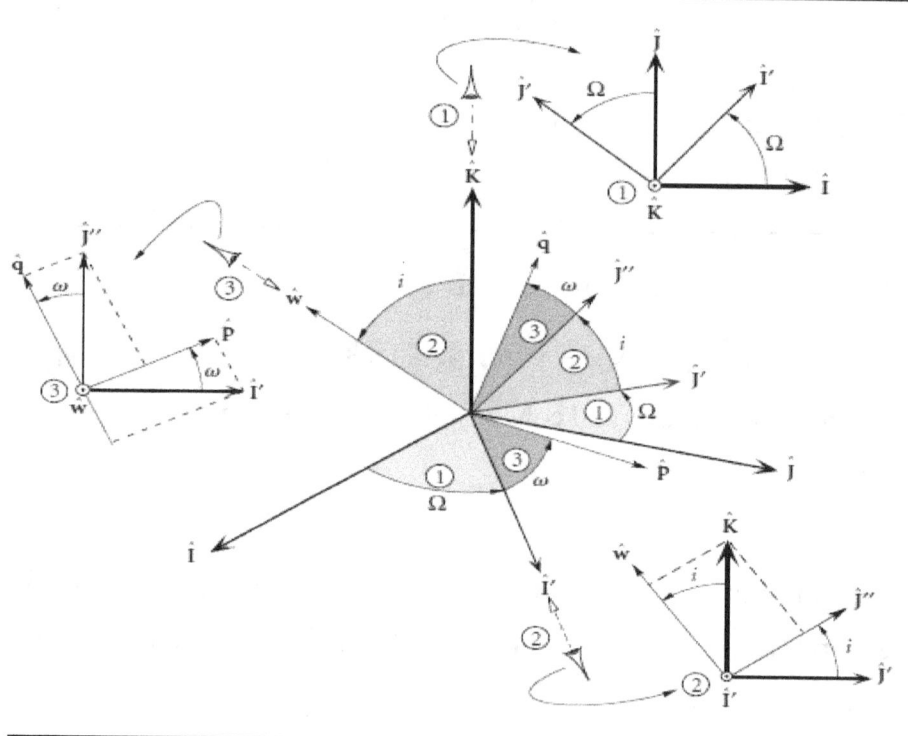

Figure 5.9: Rotation of axes

The first rotation is around the $\hat{K}$ axis, through the right ascension Ω. The orthogonal transformation matrix for this rotation is:

$$[R_3(\Omega)] = \begin{bmatrix} cos\Omega & sin\Omega & 0 \\ -sin\Omega & cos\Omega & 0 \\ 0 & 0 & 1 \end{bmatrix} \qquad (5.31)$$

Subscript on **R** refers to direction of rotation, in this case the $\hat{K}$ axis. The second rotation, is around the node line $(\hat{l})$, through the angle i required to bring the XY plane parallel to the orbital plane. The orthogonal transformation matrix is:

$$[R_1(i)] = \begin{bmatrix} 1 & 0 & 0 \\ 0 & cosi & sini \\ 0 & -sini & cosi \end{bmatrix} \qquad (5.32)$$

The third and final rotation, is in the orbital plane and rotates the unit vectors $\hat{\imath}$ and $\hat{\jmath}$ through the angle ω around the $\hat{w}$ axis so that they become aligned with $\hat{p}$ and $\hat{q}$, respectively. The transformation matrix for this rotation is:

$$[R_3(\omega)] = \begin{bmatrix} cos\omega & sin\omega & 0 \\ -sin\omega & cos\omega & 0 \\ 0 & 0 & 1 \end{bmatrix} \qquad (5.33)$$

Finally the transformation matrix $[Q]_{X\bar{x}}$ from the geocentric equatorial frame into the perifocal frame is the product of the three rotation matrices.

$$[Q]_{X\hat{x}} = [R_3(\omega)][R_1(i)][R_3(\Omega)]$$

(5.34)

The transpose of $[Q]_{X\hat{x}}$ is $[Q]_{\hat{x}X}$. It is used to transform perifocal coordinates to geocentric equatorial plane. Given the orbital elements of any planet at a given instant, the state vector of the planet can be easily calculated using equations from the perifocal frame. According to equations 2. , 2., 2. and 2. , the state vector in the perifocal frame is defined as:

$$r = \bar{x}\hat{p} + \bar{y}\hat{q} = \frac{h^2}{\mu}\frac{1}{1+e\cos\theta}(\cos\theta\hat{p} + \sin\theta\hat{q})$$

(5.35)

And

$$v = \dot{x}\hat{p} + \dot{y}\hat{q} = \frac{\mu}{h}[-\sin\theta\hat{p} + (e + \cos\theta)\hat{q}]$$

(5.36)

Utilizing the transformation matrix the perifocal frame of the planet whose state vector is to be calculated, is rotated by three angles to the perifocal frame of the Earth, the ecliptic plane. The state vectors obtained after the transformation are the position vectors of the planet relative to Earth. These are the coordinated as viewed from the ecliptic plane and distances and trajectories of the planet relative to Earth can be calculated. The position vectors obtained utilizing the transformation matrix can be defined as:

$$\{r\}_X = [Q]_{xX}\{r\}_x \qquad\qquad \{v\}_X = [Q]_{xX}\{v\}_x \qquad (5.37)$$

MATLAB ALGORITHM #2: sv_from_coe.m

This function computes the state vector (r,v) in the ecliptic plane by transforming state vector in the perfocal fram using the classical orbital elements (coe) extracted from algorithm #1.

Planet_id – As defined before.

r – Heliocentric position vector corresponding to the planet id at a given date

v – Heliocentric velocity vector corresponding to the planet id at a given date

Subfunctions required – planetary_elements, zero_to_360, kepler_E

zero_to_360 – Binds the argument to lie between 0^0 and 360^0

kepler_E – Solves Kepler's equation to solve for Eccentricity anomaly E

planetary_elements – algorithm #1

Results for state vector for earth at 1st April 2012:

```
>> global mu
>> mu = 1.327124e11;
>> deg = pi/180;
>> year1 = 2012;
>> month1 = 4;
>> % Obtain elements and rates
>> [J2000_coe1, rates1] = planetary_elements(planet_id1);
>> % Calculate Julian day
>> jd1 = 367*year1 - fix(7*(year1 + fix((month1 + 9)/12))/4) ...
+ fix(275*month1/9) + 1721013.5;
>> t01 = (jd1 - 2451545)/36525;
>> elements1 = J2000_coe1 + rates1*t01;
>> a = elements1(1);
>> e = elements1(2);
>> h = sqrt(mu*a*(1 - e^2));
>> incl = elements1(3);
>> RA = zero_to_360(elements1(4));
>> w_hat = zero_to_360(elements1(5));
>> L = zero_to_360(elements1(6));
>> w = zero_to_360(w_hat - RA);
>> M = zero_to_360((L - w_hat));
>> E = kepler_E(e, M*deg);
>> TA = zero_to_360(2*atan(sqrt((1 + e)/(1 - e))*tan(E/2))/deg);
>> [r, v] = sv_from_coe([h e RA*deg incl*deg w*deg TA*deg])

r =

  1.0e+008 *

  -1.4694   -0.2727    0.0000


v =

   4.9500  -29.4007    0.0007
```

The three components of position vector r and velocity vector v are displayed at 1st April, 2012.

5.4. Planetary Ephimeris

At this stage, we are in a position to calculated the position and velocity of any celestial body (m) in motion around the another much larger (M>>m) body due to the gravitational force of attraction. Thus, considering the Sun to be our larger body, and any other body inside the sphere of gravitational influence of the sun to by the smaller body, the trajectory of the smaller body can be determined with respect to the Sun. Or, taking the Sun to be the centre of the inertial frame of reference, we can calculate the position and velocity of any planet around it, if the orbital elements and centennial rates of the planet are known.

In order to calculate the elements and rates at a particular time instant, a system of measuring time, known as Sidereal time is utilized. In order to deduce the orbit of a satellite or celestial body from observations requires, among other things, recording the time of each observation. Sidereal time is measured by the rotation of the earth relative to the fixed stars. The time it takes for a distant star to return to its same position overhead, is one sidereal day (24 sidereal hours). Universal time (UT) is determined by the sun's passage across the Greenwich meridian, which is zero degrees terrestrial longitude. At noon UT the sun lies on the Greenwich meridian.

The Julian day number is the number of days since noon UT on 1 January 4713 BC. The origin of this time scale is placed in such a prehistoric time frame so that, we do not have to deal with positive and negative dates. The Julian day count is uniform and continuous and does not involve leap years or different numbers of days in different months. The number of days between two events is simply the difference in the Julian day numbers of the two events.

At any UT, the Julian day is given by

$$JD = J_0 + \frac{UT}{24}$$

(5.38)

One of the formulas for calculating Julian day number is (chapter 5, Reference 3):

$$J_0 = 367y - INT\left\{\frac{7\left[y + INT\left(\frac{m+9}{12}\right)\right]}{4}\right\} + INT\left(\frac{275m}{9}\right) + d + 1721013.5$$

(5.39)

Where y, m and d stand for the year, month and day ranging from:

$1901 \leq y \leq 2099$

$1 \leq m \leq 12$

$1 \leq d \leq 31$

And, INT(x) refers to only the integer portion of x, round towards zero.

Our primary aim in this section is to obtain the state vector of Earth for all the 1st day of each months from 2011 to 2099 and the state vector for Neptune for the 1st day of each month from 2021 to 3009. The solution of this problem is given in the algorithm 'ephemeris.m'.

MATLAB ALGORITHM #3: ephemeris.m

This function calculates the state vector of a planet, given the corresponding planet_id, for the 1st day of every month from the year 2011 to the year 2099.
Input variables : planet_id, Initial year.
Output variables:
R (double)(1068x3) – A 1068x3 matrix containing the $\hat{i}, \hat{j}, \hat{k}$ components of position vector in R(i,1) R(i,2) and R(i,3). Where i is the number of elapsed since initial time.
V (double)(1068x3) – Similar matrix but for the $\hat{i}, \hat{j}, \hat{k}$ components of the velocity vector.
JD – corresponding column matrix for the Julian day of each calculation.

Results:

```
Command Window                                                    →| □ ⊿ ✕
ⓘ New to MATLAB? Watch this Video, see Demos, or read Getting Started.        ✕

   >> global mu
   >> mu = 1.327124e11;
   >> planet_id1 = 3;
   >> year1 = 2011;
   >> % State vectors of Earth:
   >> [R1, V1, JD1] = ephemeris( planet_id1, year1);
   >> %Similarly for Neptune. We set the initial year to be...
   >> % 2011 + time of flight = 2021
   >> planet_id2 = 8;
   >> year2 = 2021;
   >> [R2, V2, JD2] = ephemeris( planet_id2, year2);
fx >>
```

The values for row matrices of position and velocity vectors are tabulated in excel tables and attached below:

1. Position vectors of earth –   State Vectors- Earth and Neptune\Posititon Vector-Earth.xlsx
2. Position vectors of Neptune – State Vectors- Earth and Neptune\Posititon Vector-Neptune.xlsx
3. Velocity vectors of Earth – State Vectors- Earth and Neptune\Velocity Vector-Earth.xlsx
4. Velocity vectors of Neptune – State Vectors- Earth and Neptune\Velocity Vector-Neptune.xlsx

Chapter 6: Interplanetary Spacecraft Trajectories

After calculating he position and velocity vectors of Earth and Neptune for all the months from the year 2011 to the year 2099, the next step is to launch a spacecraft from Earth to Neptune. The primary aim of this section is to present the theory required to determine the trajectory of a spacecraft launched from one planet in the solar system to another. The time of flight of the spacecraft from one planet to another is dependent on the type of propulsion system utilized by the spacecraft. As concluded earlier, we will be using ion-propulsion based propulsion system for our mission due to the enormous distances involved between Earth and Neptune. The time of flight is estimated to be around 15 years and all calculations are made based on this time of flight.

This section also presents the algorithms developed in MATLAB to calculate, at every month from years ranging from 2011 to 2099, the changes in velocity that will be required for a two- impulse maneuver from Earth to Neptune. Since the time of flight is fixed, the prime motive of any mission is to utilize as little energy in making the flight to its predetermined location. Hence the algorithm calculated the changes in velocity (delta-V) required for the two- impulse maneuver from Earth

110

to Neptune, for the time of launch corresponding to every month in years ranging from 2011 to 2099, individually and locates the month and year corresponding to minimum delta-V.

Hence the algorithm developed utilizing the theory from this section selects the best time of launch for a spacecraft from Earth to Neptune, requiring minimum energy, and also computes the orbital elements of the trajectory followed by the spacecraft in its heliocentric trajectory around the Sun to Neptune.

6.1. Sphere of Influence and the Method of Patched conics

The sun is the dominant celestial body in the solar system. It is over 1000 times more massive than the largest planet, Jupiter. The sun's gravitational pull holds all of the planets in its grasp. However, the inverse-square nature of the law of gravity causes the force of gravity, F_g, to drop off rapidly with distance r from the centre of attraction. Thus, due to the emornous distances involved between the sun and the planets, near a given planet the influence of its own gravity exceeds that of the sun. For example, at the surface of the earth, the gravitational force is over 1600 times greater than the sun's.

The sphere of influence is defined as an imaginary sphere around a celestial body up to which, its gravitation is significant as compared to the sun at a particular position in space. The ratio of gravitational force at an altitude (r) to the gravitational force at the surface of a planet (r_0), as a function of r/r_0 is depicted in figure .

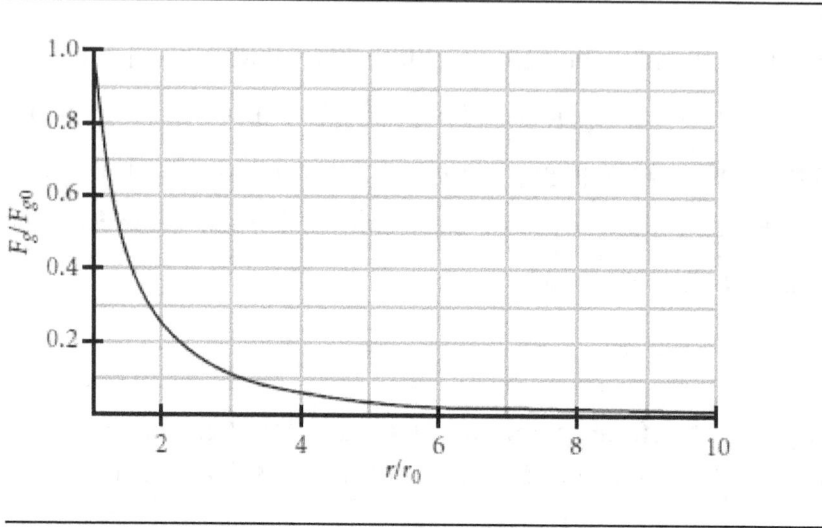

Figure 6.1: Decreasing gravitation with increasing radius

To determine the sphere of influence of any planet, we utilize the ratio of gravitational forces of the planet and sun. If $\mathbf{A}_s$ is the primary gravitational acceleration of the vehicle due to the sun, and $\mathbf{P}_p$ is the secondary or perturbing acceleration due to the planet, A_s and P_p are given by:

$$A_s = \frac{Gm_s}{R^2} \qquad P_p = \frac{Gm_p}{r^2} \qquad (5.40)$$

112

Thus, the ratio of the perturbing acceleration to the primary acceleration is:

$$\frac{P_p}{A_s} = \frac{\dfrac{Gm_p}{r^2}}{\dfrac{Gm_s}{R^2}} = \frac{m_p}{m_s}\left(\frac{R}{r}\right)^2$$

$$(5.41)$$

Near the planet, $\mathbf{a}_p$ is the primary gravitational acceleration of the vehicle due to the planet, and $\mathbf{p}_s$ is the perturbation caused by the sun. Then approximately,

$$a_p = \frac{Gm_p}{r^3}r \qquad p_s = \frac{Gm_s}{R^3}r$$

$$(5.42)$$

For motion relative to the planet, the ratio p_s/a_p is a measure of the deviation of the vehicle's orbit relative to the sun. Likewise, P_p/A_s is a measure of the planet's influence on the orbit of the vehicle relative to the sun. If,

$$\frac{p_s}{a_p} < \frac{P_p}{A_s}$$

Then, the perturbing effect of the sun on the vehicle's orbit around the planet is less than the perturbing effect of the planet on the vehicle's

113

orbit around the sun. Thus, the vehicle is therefore within the planet's sphere of influence. An equation that approximately defines the radius of the sphere of influence of a planet:

$$\frac{r_{SOI}}{R} = \left(\frac{m_p}{m_s}\right)^{2/5}$$

(5.43)

Our spaceprobe will travel from a parking orbit in the sphere of influence of Earth, out of the Earth sphere of influence and into the sphere of influence of the Sun. It will then take a heliocentric path utilizing the gravitational force of the sun to rendezvous Neptune and enter its sphere of influence.

Thus, a method of Patched conics can be deployed to calculate the state vector of the spacecraft at all times during its flight, out from the sphere of influence of the Earth, to the sphere of influence of the sun and finally into the sphere of influence of Neptune. 'Conics' refers to the fact that two-body or Keplerian orbits are conic sections with the focus at the attracting body. A very accurate assumption that can be utilized in interplanetary trajectories; we assume that when the spacecraft is outside the sphere of influence of a planet it follows an unperturbed Keplerian orbit around the sun. Because interplanetary distances are very vast, for heliocentric orbits we may neglect the size of the spheres of influence and consider them, to be just points in space coinciding

114

with the planetary centres. Within each planetary sphere of influence, however, the spacecraft travels an unperturbed Keplerian path about the planet.

To analyze a mission from planet 1(Earth) to planet 2(Neptune) using the method of patched conics, we first determine the heliocentric trajectory that will intersect the desired positions of the two planets in their orbits. This trajectory takes the spacecraft from the sphere of influence of planet 1 to that of planet 2. At the spheres of influence of the planets, computation of the heliocentric velocities of the transfer orbit is done relative to the planet to establish the velocities 'at infinity' which are then used to determine departure trajectory at planet 1 and arrival trajectory at planet 2. In this way we 'patch' together the three conics, one centred at the sun and the other two centred at the two planets.

6.2. Planetary Departure

In order to escape the gravitational pull of a planet, the spacecraft must travel a hyperbolic trajectory relative to the planet, as discussed in 'Hyperbolic orbits' in section 1.1.2. The velocity given to the spacecraft determines its hyperbolic excess velocity v_∞ relative to the planet. This is the velocity the spacecraft will possess out of the sphere of influence of the planet. If the spacecraft were to be launched in a parabolic trajectory, the spacecraft will arrive at the sphere of influence ($r = \infty$)

115

with a relative speed of zero. In that case the spacecraft remains in the same orbit as the planet and does not embark upon a heliocentric elliptical path. Figure shows the trajectory of a spacecraft departing from a parked orbit around the planet.

The spacecraft is given a boost increasing its velocity from that required to orbit at the particular orbit to greater that that required to escape the sphere of influence. The spacecraft arrives at the edge of the sphere of influence with velocity V_∞ which is the velocity required to carry the spacecraft from that planet to the sphere of influence of the targeted mission planet.

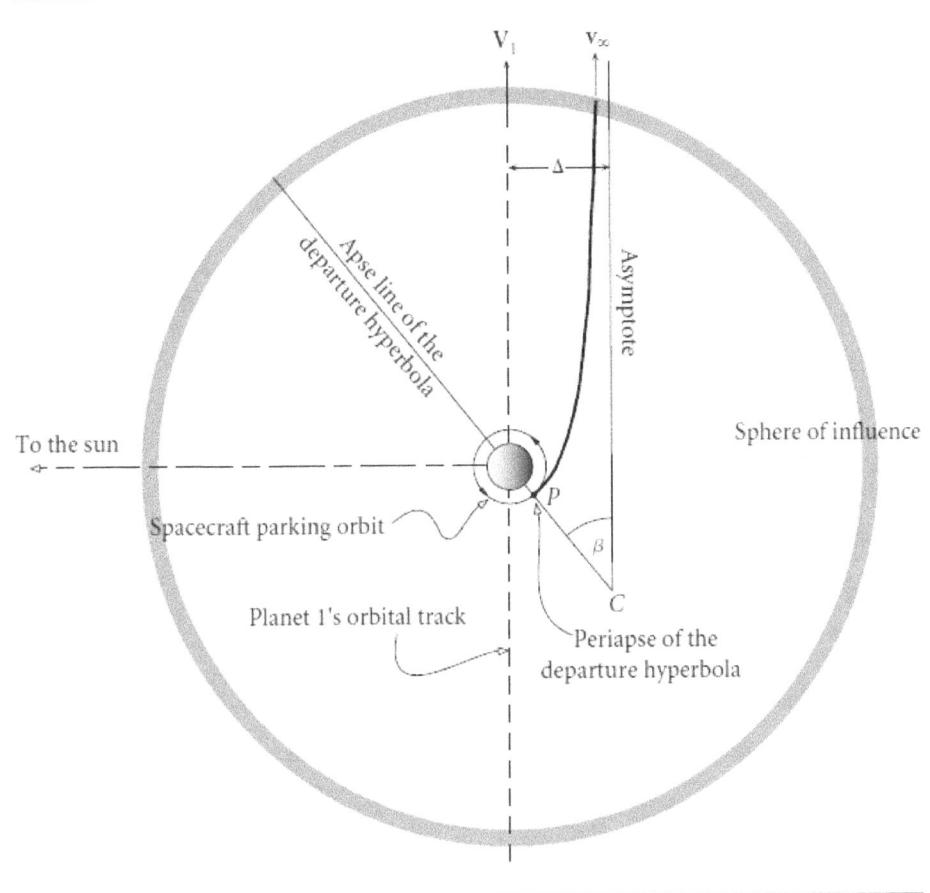

Figure 6.2: Planetary Departure

Looking at the figure, point C is the centre of the hyperbola when compared to hyperbolic trajectories in section 1.1.2. At the sphere of influence crossing, the heliocentric $V_D^{(v)}$ velocity of the spacecraft is parallel to the asymptote of the departure hyperbola as well as to the

117

planet's heliocentric velocity vector $\mathbf{V}_1$. ΔV_D is the hyperbolic excess speed of the departure hyperbola, given by,

$$v_\infty = \sqrt{\frac{\mu_{sun}}{R_1}}\left(\sqrt{\frac{2R_2}{R_1 + R_2}} - 1\right)$$

(5.44).

When a space vehicle is launched into an interplanetary trajectory from a circular parking orbit, the radius of this parking orbit equals the periapse radius r_p of the departure hyperbola. According to Equation 2.40, the periapse radius is given by

$$r = \frac{h^2}{\mu_1}\frac{1}{1 + e\cos\Theta}$$

(5.45)

where h is the angular momentum of the departure hyperbola (relative to the planet), e is the eccentricity of the hyperbola and μ_1 is the planet's gravitational parameter. The angular momentum of the hyperbola in terms of the periapsis radius r_p and the excess hyperbolic velocity is given by the relation (chapter 8, reference 3):

$$h = r_p\sqrt{v_\infty^2 + \frac{2\mu_1}{r_p}}$$

(5.46).

Since the hyperbolic excess speed is specified by the mission requirements and the rajectory to be followed, choosing a departure periapse r_p yields the parameters e and h of the departure hyperbola.

118

From the above equation we can obtain the velocity of the hyperbola at periapsis.

$$v_p = \frac{h}{r_p} = \sqrt{v_\infty^2 + \frac{2\mu_1}{r_p}}$$

(5.47)

The velocity of the spacecraft in parking orbit (considering the parking orbit to be a circular orbit around earth) is given by:

$$v_c = \sqrt{\frac{\mu_1}{r_p}}$$

(5.48)

Hence we can calculate the delta-V required to put the vehicle onto the hyperbolic departure trajectory,

$$\Delta v = v_p - v_c = v_c \left(\sqrt{2 + \left(\frac{v_\infty}{v_c}\right)^2} - 1 \right)$$

(5.49).

β is the orientation of the apse line of the hyperbola to the planet's heliocentric velocity vector. The orientation of the escape trajectory and hence the direction of velocity with which the spacecraft escapes the sphere of influence is determined by the angle β.

$$\beta = \cos^{-1}\left(\frac{1}{e}\right)$$

(5.50)

The hyperbola can be rotated about a line $A-A$ which passes through the planet's centre of mass and is parallel to v_∞. If the hyperbola is rotated about the apse line A-A, it sweeps out a surface of revolution on which lie all possible departure hyperbolas. The periapse of the hyperbola traces out a circle which, for the specified periapse radius r_p, is the locus of all possible points of injection into a departure trajectory towards the target planet.

6.3. Planetary Arrival

When a spacecraft arrives at the sphere of influence of the target planet with a hyperbolic excess velocity $v\infty$ relative to the planet from an inner planet 1 (Earth) to an outer planet 2 (Neptune), the spacecraft's heliocentric approach velocity $V_A^{(v)}$ is smaller in magnitude than that of the planet, V_2. Therefore, it crosses the forward portion of the sphere of influence, as shown in Figure . The hyperbolic excess velocity is given by:

$$v_\infty = V_2 - V_A^{(v)}$$

(5.51)

There can be three possibilities for a planetary rendezvous. If the goal is to impact the planet, the aiming radius Δ of the approach hyperbola must be such that hyperbola's periapse r_p equals the radius of the planet. If it is required to go into orbit around the planet, then Δ must be chosen so that the delta-v burn at periapse will occur at the correct altitude

120

above the planet. The delta-V burn should provide the velocity required to put the spacecraft into orbit around the planet. The relation between the periapsis radius r_p, hyperbolic excess speed, v_∞ and Δ is given by:

$$\Delta = r_p\sqrt{1 + \frac{\mu_2}{r_p v_\infty^2}} \qquad\qquad (5.52)$$

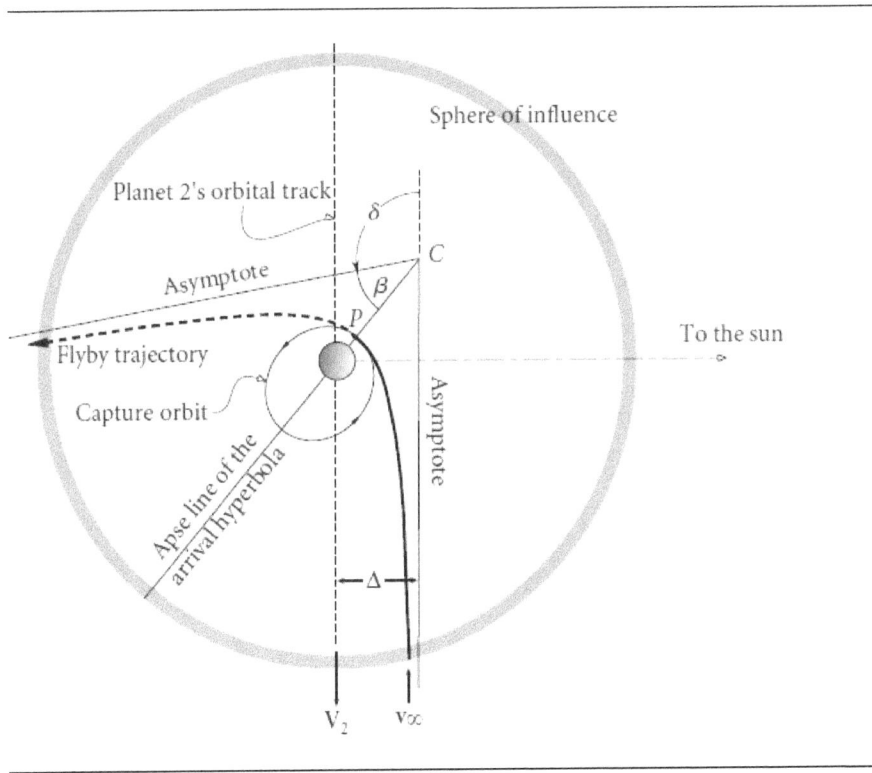

Figure 6.3: Planetary Arrival

If there is no impact with the planet and no drop into a capture orbit around the planet, then the spacecraft will simply continue past periapse on a flyby trajectory, exiting the sphere of influence with the same relative speed $v\infty$ it entered, but with the velocity vector rotated through the turn angle δ as shown in the figure. Similar to departure trajectories, the approach hyperbola does not lie in a unique plane. We can rotate the hyperbolas about a line A–A parallel to $\mathbf{v}_\infty$ and passing through the target planet's centre forming a surface of revolution.

Planetary capture:

Since our aim is to put a spacecraft into an elliptical orbit around Neptune, we will go through planetary capture. This will require a delta-v maneuver at periapse P, which is also periapse of the ellipse of capture orbit. The speed in the hyperbolic trajectory at periapse is given by

$$v_{p)hyp} = \sqrt{v_\infty^2 + \frac{2\mu_2}{r_p}}$$

(5.53)

When in an elliptical orbit around the planet, the velocity at the periapse is given by (section 1.1.2):

$$v_{p)capture} = \sqrt{\frac{\mu_2(1 + e)}{r_p}}$$

(5.54)

The required delta-V can thus be written as:

$$\Delta v = v_{p)hyp} - v_{p)capture} = \sqrt{v_\infty^2 + \frac{2\mu_2}{r_p}} - \sqrt{\frac{\mu_2(1+e)}{r_p}} \qquad (5.55)$$

Calculating the periapse radius for minimal delta-V, we obtain:

$$r_p = \frac{2\mu_2}{v_\infty^2} \frac{1-e}{1+e} \qquad (5.56)$$

Thus, from equation (above),

$$\Delta = \sqrt{\frac{2}{1-e} r_p} \qquad (5.57)$$

6.4. Orbit Determination: Lambert's Problem

In order to send a spacecraft from planet 1 to planet 2 in a specified time t_{12}, the first calculations that we require are the position and velocity vectors of the two planets. The state vector of planet 1 (R_1, V_1) at departure is to be known along with the time of departure (t_1) and the state vector of planet 2 (R_2, V_2) at arrival ($t_1 + t_{12}$). Once acquired, the method of patched conics lets us assume that the initial and final position vectors of the spacecraft are R_1 and R_2. In order to calculate the orbital elements of the transfer trajectory from planet 1 to planet 2, we need the state vectors of the spacecraft at both departure and arrival.

123

Having obtained the position vectors of the two planets from the algorithms previously described, we can solve for the values of velocity of the spacecraft at R_1 and R_2 using Lambert's problem.

Lambert's problem presents a solution to calculating the velocity of a body m in orbit around another body M, given the position vectors at any two points on the path of m around M. The velocity of the mass m around M can thus be calculated at those position vectors.

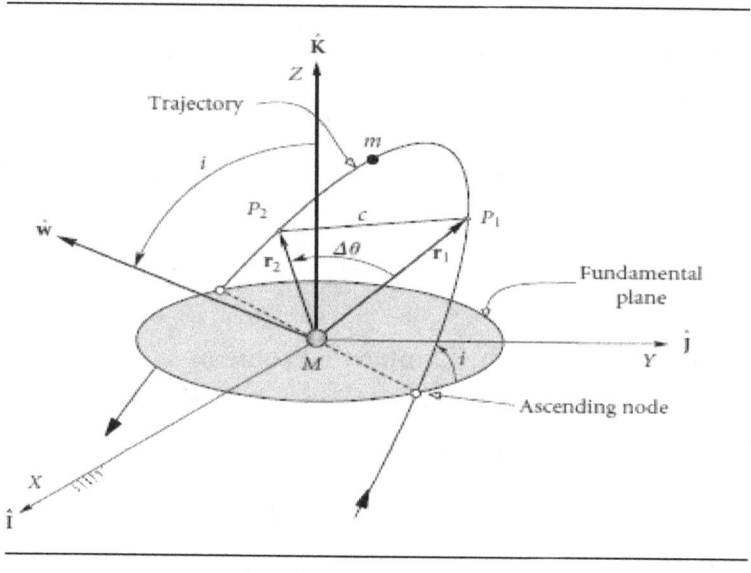

Figure 6.4: Interplanetary trajectory

Suppose we know the position vectors r_1 and r_2 of two points P_1 and P_2 on the path of mass m around mass M, as illustrated in Figure 5.3. r_1 and r_2 determine the change in the true anomaly θ,

$$cos\Delta\Theta = \frac{r_1 . r_2}{r_1 r_2} \qquad (5.58)$$

The sign of the z-component of cross product of r_1 and r_2, i.e, the sign of scalar $(r_1 \times r_2)_z$ is used to determine the correct quadrant for Θ. Two kinds of trajectories for the mass m around M are possible: prograde trajectories $(0<i<90^0)$, and retrograde trajectories $(90^0<i<180^0)$

Thus, $\Delta\Theta$ can be expressed for the following cases:

For prograde trajectory,

$$\Delta\theta = \begin{cases} cos^{-1}\left(\dfrac{r_1 . r_2}{r_1 r_2}\right) & if \ (r1 \times r2)z \geq 0 \\ 360^0 - cos^{-1}\left(\dfrac{r_1 . r_2}{r_1 r_2}\right) & if \ (r1 \times r2)z < 0 \end{cases}$$

$$(5.59)$$

For retrograde trajectory,

$$\Delta\theta = \begin{cases} cos^{-1}\left(\dfrac{r_1 . r_2}{r_1 r_2}\right) & if \ (r1 \times r2)z < 0 \\ 360^0 - cos^{-1}\left(\dfrac{r_1 . r_2}{r_1 r_2}\right) & if \ (r1 \times r2)z \geq 0 \end{cases}$$

$$(5.60)$$

If we know the time of flight Δt from P_1 to P_2, then Lambert's problem is to find the trajectory joining P_1 and P_2. The trajectory and thus, the position and velocity of any point on the path can be determined once we find $\mathbf{v}_1$. This is done through the use of Lagrange Coefficients. If the position vector $\mathbf{r}_1$ is known for a body in an orbit around another body, then the position vector, $\mathbf{r}_2$ and velocity vectors, $\mathbf{v}_1$ and $\mathbf{v}_2$ can be determined from the following equations (chapter 2, reference 3):

$$v_1 = \frac{1}{g}(\dot{g}r_2 - r_1)$$

(5.61)

And,

$$v_2 = \dot{f}r_1 + \dot{g}v_1$$

(5.62)

Where f and g are the Lagrange coefficients and $\dot{f}$ and $\dot{g}$ are their time derivatives. The Lagrange's functions and their time derivatives can be expressed in terms of the universal variable, χ (Bond and Allman(1996)).

$$f = 1 - \frac{\chi^2}{r_1}C(z) \qquad\qquad g = \Delta t - \frac{1}{\sqrt{\mu}}\chi^3 S(z)$$

(5.63)

$$\dot{f} = \frac{\sqrt{\mu}}{r_1 r_2}[zS(z) - 1] \qquad\qquad \dot{g} = 1 - \frac{\chi^2}{r_2}C(z)$$

(5.64)

where $z = \alpha\chi^2$ and is an imaginary variable. S(z) and C(z) are stumpff functions. These are a class of functions that are have an imaginary variable. They are infinitely long series. For further reference to stumpff functions refer to reference 3. These functions are iteratively solved for values of C(z) and S(z). Thus the values of the Lagrange's coefficients and their time derivatives are calculated using equations.

MATLAB ALGORITHM# 5: launch_specs.m

This function returns

1) The best time of launch from planet 1 to planet 2 by calculating the time of launch that would require the least change in velocity for the spacecraft based on a predefined time of flight from planet 1 to planet 2. The time of flight is approximated based on the propulsion system utilized by the spacecraft.

2) The orbital elements of the transfer ellipse of the spacecrafts trajectory

In this case, planet1 = Earth and planet_2 = Neptune.

Input arguments: planet_id1, planet_id2

Output arguments: year and month of launch, trajectory i.e. V1 and V2 at R1 and R2, del_V, i.e change in velocity required and coe (classical orbital elements of the transfer ellipse)

```
Command Window                                                        → □ ↗
ⓘ New to MATLAB? Watch this Video, see Demos, or read Getting Started,
  >> planet_id1 = 3;
  >> planet_id2 = 8;
  >> % Mission from Earth to Neptune.
  >> [year , month, trajectory, del_V, coe] = launch_specs(planet_id1, planet_id2);
  ------------------------------------------------------

   Departure:

   Planet: Earth
   Year : 2073
   Month : October

   Julian day: 2478450.500

   Planet position vector (km) = [1.39247e+008 -5.85006e+007 3417.38]
   Magnitude = 1.51037e+008

   Planet velocity (km/s) = [11.0542 27.3502 -0.00488263]
   Spacecraft velocity (km/s) = [15.9023 39.6659 -0.145376]
   Magnitude = 42.7351

   v-infinity at departure (km/s) = [4.84813 12.3157 -0.140493]
   Magnitude = 13.2363


   Time of flight = 3652 days
```

```
Arrival:

Planet: Neptune
Year : 2083
Month : October

Julian day: 2482102.500

Planet position vector (km) = [-2.80516e+009 3.5105e+009 -7.40323e+006]
Magnitude = 4.49362e+009

Planet velocity (km/s) = [-4.27465 -3.35592 0.167459]
Magnitude = 5.43718

Spacecraft Velocity (km/s) = [-8.12741 7.87037 -0.0142137]
Magnitude = 11.3136

v-infinity at arrival (km/s) = [-3.85276 11.2263 -0.181673]
Magnitude = 11.8704


Orbital elements of flight trajectory:

Angular momentum (km^2/s) = 6.45368e+009
Eccentricity = 1.07843
Right ascension of the ascending node (deg) = 157.593
Inclination to the ecliptic (deg) = 0.194918
Argument of perihelion (deg) = 181.435
True anomaly at departure (deg) = 358.184
True anomaly at arrival (deg) = 149.6
Semimajor axis (km) = -1.9253e+009>>
```

CONCLUSION

The first, second and third chapters were just a basic study into robotic space probes. It defined the types of missions that are required to be performed by various kinds of spaceprobes. It defined a general outline and a vague direction for our heading for a planetary rendezvous with Neptune.

Chapters 4 and 5 discuss the design of the spacecraft, which will be feasible to send to Neptune. The design is discussed based on the subsystems and the scientific instruments that will be used in a mission similar to Voyager 2. It also sheds light on the problems faced while preparing for a mission to Neptune such as that of propulsion and power source. It was concluded that the three-axis stabilized system will use celestial or gyro referenced attitude control to maintain pointing of the high-gain antennas toward Earth. The spacecraft would carry a Flight Data Subsystem (FDS) and a single 8-track Digital tape recorder (DTR) to provide the data handling functions. It would require a command computer subsystem (CCS) and an attitude and articulation control subsystem that would control orientation of the spacecraft.

The spacecraft would suport on-board, a host of scientific instruments. Investigator teams such as plasma science (PLS), low-energy charged particles (LECP), Cosmic Ray Subsystem (CRS), Magnetometer (MAG) and Plasma Wave Subsystem (PWS).

Due to the large distances involved in the travel between Earth and Neptune, a study of emerging propulsion techniques makes ion propulsion the most appropriate choice for traversing distances as far as the edge of the solar system and beyond. The ion propulsion system deployed would be similar to the DS-1 ion engine with a specific impulse of nearly 8000 seconds.

The power subsystem utilized in the space probe was determined to be a controlled nuclear- thermoelectric source. Neptune receives only 3percent the radiation received by Jupiter. Thus, a nuclear thermoelectric source becomes an obvious choice for the power subsystem due to the scarcity of sunlight intensity at far off distances in the solar system.

Based on the choice of propulsion, the time period it would require for a spacecraft to reach Neptune was determined to be approximately about ten years. Chapters 5 and 6 provide the basis for trajectory calculation and preliminary orbit determination of the planets Earth and Neptune and the spacecraft in heliocentric paths around the Sun. The algorithms developed with the use of MATLAB were used to calculated the time period of launch from 2011 to 2099 that would require the minimum delta-V or change in velocity to leave the sphere of influence of Earth, travel through space in heliocentric path around the Sun and enter Neptune's sphere of influence.

The following is a summary of the analysis:

Departure Details:

Planet: Earth

Recommended date of launch : October, 2073

Julian day: 2478450.500

Planet position vector (km) = $[1.39247\text{e}+008\hat{i},$ $-5.85006\text{e}+007\hat{j},$

  $3417.38\hat{k}$]

Magnitude = 1.51037e+008

Planet velocity (km/s) = $[11.0542\hat{i},$ $27.3502\hat{j},$ -

$0.00488263\hat{k}]$

Spacecraft velocity (km/s) = $[15.9023\hat{i},$ $39.6659\hat{j},$ $-0.145376\hat{k}]$

Magnitude = 42.7351

v-infinity at departure (km/s) = $[4.84813\hat{i}$ $12.3157\hat{j}$ $-0.140493\hat{k}]$

Magnitude = 13.2363

Time of flight = 3652 days

Arrival:

Planet: Neptune

Year : 2083

Month : October

Julian day: 2482102.500

Planet position vector (km) = [-2.80516e+009$\hat{i}$,   3.5105e+009$\hat{j}$,

     -7.40323e+006$\hat{k}$]

Magnitude = 4.49362e+009

Planet velocity (km/s) = [-4.27465$\hat{i}$,   -3.35592$\hat{j}$,

    0.167459$\hat{k}$]

Magnitude = 5.43718

Spacecraft Velocity (km/s) = [-8.12741$\hat{i}$,   7.87037$\hat{j}$,   -0.0142137
$\hat{k}$]

Magnitude = 11.3136

v-infinity at arrival (km/s) = [-3.85276$\hat{i}$,   11.2263$\hat{j}$,   -0.181673$\hat{k}$]

Magnitude = 11.8704

Orbital elements of flight trajectory:

Angular momentum (km^2/s) = 6.45368e+009

Eccentricity = 1.07843

Right ascension of the ascending node (deg) = 157.593

Inclination to the ecliptic (deg) = 0.194918

Argument of perihelion (deg) = 181.435

True anomaly at departure (deg) = 358.184

True anomaly at arrival (deg) = 149.6

Semimajor axis (km) = -1.9253e+009

APPENDIX:

MATLAB ALGORITHMS

---

1. MATLAB ALGORITHM #1: planetary_elements.m

```
%-------------------------------------------------------------------------
function [ J2000_coe, rates ] = planetary_elements( planet_id )
```

```
%-------------------------------------------------------------------------
% This function extracts a planet's J2000 orbital elements and
% centennial rates
%
% planet_id - 1 to 9, for Mercury to Pluto
%
% J2000_elements - 9 by 6 matrix of J2000 orbital elements for
% the nine planets Mercury through Pluto.
% cent_rates - 9 by 6 matrix of the rates of change of the
% J2000_elements per Julian century (Cy).

J2000_elements = ...
[ 0.38709893 0.20563069 7.00487 48.33167 77.45645 252.25084
0.72333199 0.00677323 3.39471 76.68069 131.53298 181.97973
1.00000011 0.01671022 0.00005 -11.26064 102.94719 100.46435
1.52366231 0.09341233 1.85061 49.57854 336.04084 355.45332
5.20336301 0.04839266 1.30530 100.55615 14.75385 34.40438
9.53707032 0.05415060 2.48446 113.71504 92.43194 49.94432
19.19126393 0.04716771 0.76986 74.22988 170.96424 313.23218
30.06896348 0.00858587 1.76917 131.72169 44.97135 304.88003
39.48168677 0.24880766 17.14175 110.30347 224.06676 238.92881];
cent_rates = ...
[ 0.00000066 0.00002527 -23.51 -446.30 573.57 538101628.29
0.00000092 -0.00004938 -2.86 -996.89 -108.80 210664136.06
-0.00000005 -0.00003804 -46.94 -18228.25 1198.28 129597740.63
-0.00007221 0.00011902 -25.47 -1020.19 1560.78 68905103.78
0.00060737 -0.00012880 -4.15 1217.17 839.93 10925078.35
-0.00301530 -0.00036762 6.11 -1591.05 -1948.89 4401052.95
0.00152025 -0.00019150 -2.09 -1681.4 1312.56 1542547.79
-0.00125196 0.00002514 -3.64 -151.25 -844.43 786449.21
-0.00076912 0.00006465 11.07 -37.33 -132.25 522747.90];
```

```
J2000_coe = J2000_elements(planet_id,:);
rates = cent_rates(planet_id,:);

% Convert from AU to km:
au = 149597871;
J2000_coe(1) = J2000_coe(1)*au;
rates(1) = rates(1)*au;
% Convert from arcseconds to fractions of a degree:
rates(3:6) = rates(3:6)/3600;

return

end
```

---

2. MATLAB ALGORITHM #2: sv_from_coe.m

```
%------------------------------------------------------------------------
function [ r, v ] = sv_from_coe( coe )
%------------------------------------------------------------------------
%This function computes the state vector (r,v) from the
% classical orbital elements (coe).

% R3_w - Rotation matrix about the z-axis through the angle w
% R1_i - Rotation matrix about the x-axis through the angle i
% R3_W - Rotation matrix about the z-axis through the angle RA
% Q_pX - Matrix of the transformation from perifocal to
% rp - position vector in the perifocal frame (km)
% vp - velocity vector in the perifocal frame (km/s)
% r - position vector in the geocentric equatorial frame(km)
% v - velocity vector in the geocentric equatorial frame(km/s)
```

136

```
global mu
h = coe(1);
e = coe(2);
RA = coe(3);
incl = coe(4);
w = coe(5);
TA = coe(6);

rp = (h^2/mu) * (1/(1 + e*cos(TA))) * (cos(TA)*[1;0;0] ...
+ sin(TA)*[0;1;0]);
vp = (mu/h) * (-sin(TA)*[1;0;0] + (e + cos(TA))*[0;1;0]);

R3_W = [ cos(RA) sin(RA) 0
    -sin(RA) cos(RA) 0
      0    0   1];

R1_i = [1    0    0
    0  cos(incl)  sin(incl)
    0  -sin(incl) cos(incl)];

R3_w = [ cos(w) sin(w) 0
    -sin(w) cos(w) 0
      0   0   1];

Q_pX = R3_W'*R1_i'*R3_w';

r = Q_pX*rp;
v = Q_pX*vp;

% Convert r and v into row vectors:
```

```
r = r';
v = v';
```

end

_____

3. MATLAB ALGORITHM #3 : ephimeris.m

```
%----------------------------------------------------------------------------
function [ R, V, JD ] = ephemeris( planet_id, year )
%----------------------------------------------------------------------------
% This function calculates the orbital elements and the state
% vector of a planet at the 1st day of each month
% from 2011 to 2099
%
% mu - gravitational parameter of the sun (km^3/s^2)
% deg - conversion factor between degrees and radians
% pi - 3.1415926...
%
% coe - vector of heliocentric orbital elements
% [h e RA incl w TA a w_hat L M E],
% where
% h = angular momentum (km^2/s)
% e = eccentricity
% RA = right ascension (deg)
% incl = inclination (deg)
% w = argument of perihelion (deg)
% TA = true anomaly (deg)
% a = semimajor axis (km)
```

% w_hat = longitude of perihelion
% (= RA + w) (deg)
% L = mean longitude (= w_hat + M) (deg)
% M = mean anomaly (deg)
% E = eccentric anomaly (deg)
%
% planet_id - planet identifier:
% 1 = Mercury
% 2 = Venus
% 3 = Earth
% 4 = Mars
% 5 = Jupiter
% 6 = Saturn
% 7 = Uranus
% 8 = Neptune
% 9 = Pluto
%
% year - range: 1901 - 2099
% month - range: 1 - 12
% day - range: 1 - 31
% hour - range: 0 - 23
% minute - range: 0 - 60
% second - range: 0 - 60
%
% j0 - Julian day number of the date at 0 hr UT
% ut - universal time in fractions of a day
% jd - julian day number of the date and time
%
% J2000_coe - row vector of J2000 orbital elements
% rates - row vector of Julian centennial rates
% t0 - Julian centuries between J2000 and jd

```
% elements - orbital elements at jd
%
% r - heliocentric position vector
% v - heliocentric velocity vector
%
% User subfunctions required: planetary_elements, zero_to_360,
kepler_E, sv_from_coe


global mu
deg = pi/180;

R = zeros(1068,3);
V = zeros(1068, 3);
JD = zeros(1068,1);
[J2000_coe, rates] = planetary_elements(planet_id);
n = 1;
month = 0;
for ii = 1:89;
   for jj = 1:12
jd = 367*year - fix(7*(year + fix((month + 9)/12))/4) ...
+ fix(275*month/9) + 1721013.5;

t0 = (jd - 2451545)/36525;

elements = J2000_coe + rates*t0;
a = elements(1);
e = elements(2);

h = sqrt(mu*a*(1 - e^2));
```

```
% Reduce the angular elements to within the range 0 - 360 degrees:
incl = elements(3);
RA = zero_to_360(elements(4));
w_hat = zero_to_360(elements(5));
L = zero_to_360(elements(6));
w = zero_to_360(w_hat - RA);
M = zero_to_360((L - w_hat));

% Algorithm kepler_E
E = kepler_E(e, M*deg);

% True anomaly in degrees
TA = zero_to_360(2*atan(sqrt((1 + e)/(1 - e))*tan(E/2))/deg);


% Algorithm sv_from_coe (all angles in radians)
[r, v] = sv_from_coe([h e RA*deg incl*deg w*deg TA*deg]);

R(n,:) = [r(1), r(2), r(3)];
V(n,:) = [v(1), v(2), v(3)];
JD(n,1) = jd;
n = n+1;
month = month+1;


    end

    month = 1;
    year = year+1;

end
```

```
return
end
```

_____

Sub-functions Used in the main body:

3.1.   zero_to_360.m

```
function [ y ] = zero_to_360( x )
%This function reduces an angle to lie in the range
% 0 - 360 degrees.
%
% x - the original angle in degrees
% y - the angle reduced to the range 0 - 360 degrees
 if x >= 360
x = x - fix(x/360)*360;
elseif x < 0
x = x - (fix(x/360) - 1)*360;
end
y = x;
return
end
```

3.2.   kepler_E.m

```
function [ E ] = kepler_E( e, M )
%This function uses Newton's method to solve Kepler's
% equation E - e*sin(E) = M
```

```
%
% E - eccentric anomaly (radians)
% e - eccentricity, passed from the calling program
% M - mean anomaly (radians), passed from the calling program

error = 1.e-8;
% Starting value for E
if M < pi
E = M + e/2;
else
E = M - e/2;
end

ratio = 1;
while abs(ratio) > error
ratio = (E - e*sin(E) - M)/(1 - e*cos(E));
E = E - ratio;
end

end
```

---

4. MATLAB ALGORITHM #4 : lambert.m

```
%-------------------------------------------------------------------------
function [ v1, v2 ] = lambert( R1, R2, t, string )
%-------------------------------------------------------------------------
% This function solves Lambert's problem.
```
143

```
%
% R1, R2 - initial and final position vectors (km)
% r1, r2 - magnitudes of R1 and R2
% t - the time of flight from R1 to R2
% V1, V2 - initial and final velocity vectors (km/s)
% theta - angle between R1 and R2
% string - 'pro' if the orbit is prograde
% 'retro' if the orbit is retrograde
% y(z) - a function of z
% F(z,t) - a function of the variable z and constant t,
% f, g - Lagrange coefficients
% C(z), S(z) - Stumpff functions


global mu
global r1 r2 A
%...Magnitudes of R1 and R2:
r1 = norm(R1);
r2 = norm(R2);
c12 = cross(R1, R2);

theta = acos(dot(R1,R2)/r1/r2);
%...Determine whether the orbit is prograde or retrograde:
if strcmp(string, 'pro')
if c12(3) <= 0
theta = 2*pi - theta;
end
elseif strcmp(string,'retro')
if c12(3) >= 0
theta = 2*pi - theta;
end
```

```
else
string = 'pro'
fprintf('\n ** Prograde trajectory assumed.\n')
end


A = sin(theta)*sqrt(r1*r2/(1 - cos(theta)));

%...Determine approximately where F(z,t) changes sign, and
%...use that value of z as the starting value
z = -100;
while F(z,t) < 0
z = z + 0.1;
end
%...Set an error tolerance and a limit on the number of iterations:
tol = 1.e-8;
nmax = 5000;
%...Iterate on Equation 5.45 until z is determined to within
%...the error tolerance:
ratio = 1;
n =0;
while (abs(ratio) > tol) & (n <= nmax)
n = n + 1;
ratio = F(z,t)/dFdz(z);
z = z - ratio;
end
%...Report if the maximum number of iterations is exceeded:
if n >= nmax
fprintf('\n\n **Number of iterations exceeds')
fprintf(' %g \n\n ', nmax)
end
```

```
f = 1 - y(z)/r1;
g = A*sqrt(y(z)/mu);
gdot = 1 - y(z)/r2;
v1 = (1/g)*(R2 - f*R1);
v2 = (1/g)*(gdot*R2 - R1);
return
end
```

Subfunctions used in the main body:

4.1.   y.m

```
%------------------------------------------------------------
function [ dum ] = y( z )
%------------------------------------------------------------
global r1 r2 A
dum = r1 + r2 + A*(z*S(z) - 1)/sqrt(C(z));
return
end
```

4.2.   F.m

```
%------------------------------------------------------------
function f = F( z, t )
%------------------------------------------------------------
 global mu A
f = (y(z)/C(z))^1.5*S(z) + A*sqrt(y(z)) - sqrt(mu)*t;
return
end
```

4.3. dFdz.m

```
%--------------------------------------------------------------------------
function [ dum ] = dFdz( z )
%--------------------------------------------------------------------------
global A
if z == 0
dum = sqrt(2)/40*y(0)^1.5 + A/8*(sqrt(y(0)) ...
+ A*sqrt(1/2/y(0)));
else
dum = (y(z)/C(z))^1.5*(1/2/z*(C(z) - 3*S(z)/2/C(z)) ...
+ 3*S(z)^2/4/C(z)) ...
+ A/8*(3*S(z)/C(z)*sqrt(y(z)) ...
+ A*sqrt(C(z)/y(z)));
end
return

end
```

4.4. StumpC

```
%--------------------------------------------------------------------------
function [ c ] = stumpC( z )
%--------------------------------------------------------------------------
% This function evaluates the Stumpff function C(z)
 % z - input argument
% c - value of C(z)
 if z > 0
c = (1 - cos(sqrt(z)))/z;
elseif z < 0
```

```
c = (cosh(sqrt(-z)) - 1)/(-z);
else
c = 1/2;
end

end
```

4.5.   StumpS

```
%----------------------------------------------------------------
function [ s ] = stumpS( z )
%----------------------------------------------------------------
% This function evaluates the Stumpff function S(z) according
% z - input argument
% s - value of S(z)
 if z > 0
s = (sqrt(z) - sin(sqrt(z)))/(sqrt(z))^3;
elseif z < 0
s = (sinh(sqrt(-z)) - sqrt(-z))/(sqrt(-z))^3;
else
s = 1/6;
end

end
```

---

5. MATLAB ALGORITH #5 : launch_specs.m

```
%----------------------------------------------------------------
---------------------
function [ year1, month1, trajectory, del_V, coe ] = launch_specs(
planet_id1, planet_id2)
%----------------------------------------------------------------
---------------------
% This function returns:
% 1) The best time of launch from planet 1 to planet 2
% by calculating the time of launch that would require the
% least change in velocity for the spacecraft based on
% a predefined time of flight from planet 1 to
% planet 2. The time of flight is approximated based
% on the propulsion system utilized by the spacecraft.
% 2) The orbital elements of the transfer ellipse
% of the spacecrafts trajectory
%
% mu - gravitational parameter of the sun (km^3/s^2)
%
% planet_id - planet identifier:
% 1 = Mercury
% 2 = Venus
% 3 = Earth
% 4 = Mars
% 5 = Jupiter
% 6 = Saturn
% 7 = Uranus
% 8 = Neptune
% 9 = Pluto
% year_depart(2011)- Initial year of departure for iteration
% year_arrive(2021)- Final year of departure for iteration
% Time of flight assumed to be 10 years for Earth to Neptune
```

% year - range: 2011 - 2099
% month - range: 1 - 12
%
% jd1, jd2 - Julian day numbers at departure and arrival
% tof - time of flight from planet 1 to planet 2 (s)
%
%
% depart - [planet_id, year, month] at departure
% arrive - [planet_id, year, month] at arrival
% planet1 - [Rp1, Vp1, jd1]
% planet2 - [Rp2, Vp2, jd2]
% trajectory - [V1, V2]
% Rp1, Vp1 - state vector of planet 1 at departure (km, km/s)
% Rp2, Vp2 - state vector of planet 2 at arrival (km, km/s)
% R1, V1 - heliocentric state vector of spacecraft at
% departure (km, km/s)
% R2, V2 - heliocentric state vector of spacecraft at
% arrival (km, km/s)
% year1, month1 - Most efficient time to launch
% trajectory - Matrix containing Spacecraft velocity at departure
% and arrival from 2011 to 2099
%
% coe - Orbital elements of most efficient spacecraft trajectory
% coe = [h e RA incl w TA a]
% h - the magnitude of H (km^2/s)
% incl - inclination of the orbit (rad)
% RA - right ascension of the ascending node (rad)
% e - eccentricity (magnitude of E)
% w - argument of perigee (rad)
% TA - true anomaly (rad)
% a - semimajor axis (km)

```
% User subfunctions required: coe_from_sv , year_month_planet,
interplanetary
%------------------------------------------------------------------------
---------
global mu
deg = pi/180;

year_depart = 2011;
year_arrive = 2021;
depart = [planet_id1, year_depart];
arrive = [planet_id2, year_arrive];

%...Obtain matrices for position and velocity vector of planet1
% and planet 2, and velocity of spacecraft at departure and arrival
[planet1, planet2, trajectory] = interplanetary(depart, arrive);

R1 = planet1(:, 1:3);
Vp1 = planet1(:, 4:6);
jd1 = planet1(:,7);

R2 = planet2(:, 1:3);
Vp2 = planet2(:, 4:6);
jd2 = planet2(:,7);

%...Row matrix for times of flight
tof = jd2 - jd1;

V1 = trajectory(:,1:3);
V2 = trajectory(:,4:6);
```

```
%...Row matrices for delV required at planet1(vinf1) and planet
2(vinf2)
vinf1 = V1 - Vp1;
vinf2 = V2 - Vp2;


%....Obtain minimum del_V
min = norm(vinf1(1,:)) + norm(vinf2(1,:));
del_V = zeros(1068,1);


for i = 1:1068
del_V(i) = norm(vinf1(i,:)) + norm(vinf2(i,:));
   if del_V(i) < min
     min = del_V(i);
     l = i;
   end
end


%...Obtain year and month corresponding to min del_V
[ year1, month1, planet1 ] = year_month_planet( l, planet_id1 );
[ year2, month2, planet2 ] = year_month_planet((l+30*12), planet_id2
);


%...Obtain orbital elements of the most efficient spacecraft trajectory
coe = coe_from_sv(R1(l,:), V1(l,:));
coe2 = coe_from_sv(R2(l,:), V2(l,:));


fprintf('--------------------------------------------------')
fprintf('\n\n Departure:\n');
fprintf('\n Planet: %s', planet1)
fprintf('\n Year : %g', year1)
fprintf('\n Month : %s', month1)
```

```
fprintf('\n\n Julian day: %11.3f\n', jd1(l))
fprintf('\n Planet position vector (km) = [%g %g %g]', ...
R1(l,1), R1(l,2), R1(l,3))
fprintf('\n Magnitude = %g\n', norm(R1(l,:)))
fprintf('\n Planet velocity (km/s) = [%g %g %g]', ...
Vp1(l,1), Vp1(l,2), Vp1(l,3))
fprintf('\n Spacecraft velocity (km/s) = [%g %g %g]', ...
V1(l,1), V1(l,2), V1(l,3))
fprintf('\n Magnitude = %g\n', norm(V1(l,:)))
fprintf('\n v-infinity at departure (km/s) = [%g %g %g]', ...
vinf1(l,1), vinf1(l,2), vinf1(l,3))
fprintf('\n Magnitude = %g\n', norm(vinf1(l,:)))
fprintf('\n\n Time of flight = %g days\n', tof(l))


fprintf('\n\n Arrival:\n');
fprintf('\n Planet: %s', planet2)
fprintf('\n Year : %g', year2)
fprintf('\n Month : %s', month2)
fprintf('\n\n Julian day: %11.3f\n', jd2(l))
fprintf('\n Planet position vector (km) = [%g %g %g]', ...
R2(l,1), R2(l,2), R2(l,3))
fprintf('\n Magnitude = %g\n', norm(R2(l,:)))
fprintf('\n Planet velocity (km/s) = [%g %g %g]', ...
Vp2(l,1), Vp2(l,2), Vp2(l,3))
fprintf('\n Magnitude = %g\n', norm(Vp2(l,:)))
fprintf('\n Spacecraft Velocity (km/s) = [%g %g %g]', ...
V2(l,1), V2(l,2), V2(l,3))
fprintf('\n Magnitude = %g\n', norm(V2(l,:)))
fprintf('\n v-infinity at arrival (km/s) = [%g %g %g]', ...
vinf2(l,1), vinf2(l,2), vinf2(l,3))
```

```
fprintf('\n Magnitude = %g', norm(vinf2(1,:)))
```

```
fprintf('\n\n\n Orbital elements of flight trajectory:\n')
fprintf('\n Angular momentum (km^2/s) = %g', coe(1))
fprintf('\n Eccentricity = %g', coe(2))
fprintf('\n Right ascension of the ascending node')
fprintf(' (deg) = %g', coe(3)/deg)
fprintf('\n Inclination to the ecliptic (deg) = %g', ...
coe(4)/deg)
fprintf('\n Argument of perihelion (deg) = %g', ...
coe(5)/deg)
fprintf('\n True anomaly at departure (deg) = %g', ...
coe(6)/deg)
fprintf('\n True anomaly at arrival (deg) = %g\n', ...
coe2(6)/deg)
fprintf('\n Semimajor axis (km) = %g', coe(7))
```

```
end
```

_____

Subfunctions used in the main body:

    5.1.   coe_from_sv.m
```
%-----------------------------------------------------------------
function [ coe ] = coe_from_sv( R,V )
%-----------------------------------------------------------------
```

```
% This function computes the classical orbital elements (coe)
% from the state vector (R,V).
%
```

154

% R - position vector in the geocentric equatorial frame(km)
% V - velocity vector in the geocentric equatorial frame(km)
% vr - radial velocity component (km/s)
% H - the angular momentum vector (km^2/s)
% h - the magnitude of H (km^2/s)
% incl - inclination of the orbit (rad)
% N - the node line vector (km^2/s)
% n - the magnitude of N
% RA - right ascension of the ascending node (rad)
% E - eccentricity vector
% e - eccentricity (magnitude of E)
% eps - a small number below which the eccentricity is
% considered to be zero
% w - argument of perigee (rad)
% TA - true anomaly (rad)
% a - semimajor axis (km)
% coe - vector of orbital elements [h e RA incl w TA a]

```
global mu;
eps = 1.e-10;
r = norm(R);
v = norm(V);
vr = dot(R,V)/r;
H = cross(R,V);
h = norm(H);
incl = acos(H(3)/h);
N = cross([0 0 1],H);
n = norm(N);

if n ~= 0
RA = acos(N(1)/n);
```

```
if N(2) < 0
RA = 2*pi - RA;
end
else
RA = 0;
end

E = 1/mu*((v^2 - mu/r)*R - r*vr*V);
e = norm(E);

if n ~= 0
if e > eps
w = acos(dot(N,E)/n/e);
if E(3) < 0
w = 2*pi - w;
end
else
w = 0;
end
else
w = 0;
end

if e > eps
TA = acos(dot(E,R)/e/r);
if vr < 0
TA = 2*pi - TA;
end
else
cp = cross(N,R);
if cp(3) >= 0
```

```
TA = acos(dot(N,R)/n/r);
else
TA = 2*pi - acos(dot(N,R)/n/r);
end
end
%... (a < 0 for a hyperbola):
a = h^2/mu/(1 - e^2);
coe = [h e RA incl w TA a];
 end
```

5.2.   year_month_planet.m

```
%-------------------------------------------------------------------------
function [ year, month, planet ] = year_month_planet( l, planet_id )
%-------------------------------------------------------------------------
%YEAR_MONTH Summary of this function goes here
%   Detailed explanation goes here

y = fix(l/12);
m = rem(l, 12);

year = 2011+ y;
switch (m+1)
   case 1
      month = 'January';
   case 2
      month = 'February';
   case 3
      month = 'March';
   case 4
```

```
      month = 'April';
   case 5
      month = 'May';
   case 6
      month = 'June';
   case 7
      month = 'July';
   case 8
      month = 'August';
   case 9
      month = 'September';
   case 10
      month = 'October';
   case 11
      month = 'Novermber';
   case 12
      month = 'December';
end

switch planet_id
   case 1
      planet = 'Mercury';
   case 2
      planet = 'Venus';
   case 3
      planet =  'Earth';
   case 4
      planet = 'Mars';
   case 5
      planet = 'Jupiter';
   case 6
```

```matlab
      planet = 'Saturn';
   case 7
      planet = 'Uranus';
   case 8
      planet = 'Neptune';
   case 9
      planet = 'Pluto';

end

end
```

---

6. MATLAB ALGORITHM #6: interplanetary.m

```matlab
%-------------------------------------------------------------------------------
function [ planet1, planet2, trajectory ] = interplanetary( depart, arrive )
% -------------------------------------------------------------------------------
%
% This function determines the spacecraft trajectory from the
% sphere of influence of planet 1 to that of planet 2
% for every month from year 2011 to year 2099

% mu - gravitational parameter of the sun (km^3/s^2)
%
% planet_id - planet identifier:
% 1 = Mercury
% 2 = Venus
% 3 = Earth
```

```
% 4 = Mars
% 5 = Jupiter
% 6 = Saturn
% 7 = Uranus
% 8 = Neptune
% 9 = Pluto
%
% year - range: 2011 - 2099
% month - range: 1 - 12
%
% jd1, jd2 - Julian day numbers at departure and arrival
% tof - time of flight from planet 1 to planet 2 (s)
%
% Rp1, Vp1 - state vector of planet 1 at departure (km, km/s)
% Rp2, Vp2 - state vector of planet 2 at arrival (km, km/s)
% R1, V1 - heliocentric state vector of spacecraft at
% departure (km, km/s)
% R2, V2 - heliocentric state vector of spacecraft at
% arrival (km, km/s)
%
% depart - [planet_id, year, month] at departure
% arrive - [planet_id, year, month] at arrival
% planet1 - [Rp1, Vp1, jd1]
% planet2 - [Rp2, Vp2, jd2]
% trajectory - [V1, V2]
% Subfunctions used in the main body: lambert.m
%----------------------------------------------------------------

global mu
planet_id1 = depart(1);
year1 = depart(2);
```

```
[ Rp1, Vp1, JD1 ] = ephemeris( planet_id1, year1 );


planet_id2 = arrive(1);
year2 = arrive(2);
[ Rp2, Vp2, JD2 ] = ephemeris( planet_id2, year2 );

%...Patched conic assumption:
R1 = Rp1;
R2 = Rp2;

%... Solve lamberts problem to calcutate Spacecraft trajectory
% from planet 1 to planet 2 for years 2011 through 2099
V1 = zeros(468,3);
V2 = zeros(468,3);
MO = zeros(468);
y = zeros(468);

for i = 1:468;
tof = (JD2(i,:) - JD1(i,:))*24*3600;
[v1, v2] = lambert(R1(i,:), R2(i,:), tof, 'pro');
V1(i,:) = v1;
V2(i,:) = v2;
end
planet1 = [Rp1, Vp1, JD1];
planet2 = [Rp2, Vp2, JD2];
trajectory = [V1, V2];

end
```

References:

[1] **Joseph, Angelo A.** *Robot Spacecrafts* Infobase Publishing, 2007

[2] **Damon, Thomas D**. *Introduction to Space: The Science of Spaceflight.* 3d ed
> Malabar,Fla.: Krieger Publishing Co., 2000.

[3] **Curtis, Howard D.**, *Orbital mechanics for engineering students*

[4] **Heppenheimer, Thomas A.** *Countdown: A History of Space Flight.* New York:
> John Wiley and Sons, 1997.

Planetary data for Neptune

*Time required for the planet to return to the same position in the sky relative to the Sun as seen from Earth.

**Calculated for the altitude at which 1 bar of atmospheric pressure is exerted.

Mean Distance From Sun	4,498,396,000 Km (30.1 AU)
Eccentricity Of Orbit	0.0086
Inclination Of Orbit To Ecliptic	1.77°
Neptunian Year (Sidereal Period Of Revolution)	164.79 Earth Years
Visual Magnitude At Mean Opposition	7.8
Mean Synodic Period*	367.49 Earth Days
Mean Orbital Velocity	5.43 Km/Sec
Equatorial Radius**	24,764 Km
Polar Radius**	24,340 Km
Mass	1.02×10^{26} Kg
Mean Density	1.64 G/Cm^3
Gravity**	1,115 Cm/Sec^2
Escape Velocity**	23.6 Km/Sec
Rotation Period (Magnetic Field)	16 Hr 7 Min
Inclination Of Equator To Orbit	28.3°
Magnetic Field Strength At Equator (Mean)	0.14 Gauss
Tilt Angle Of Magnetic Axis	46.8°

Planetary data for Neptune

Offset Of Magnetic Axis	0.55 Of Neptune's Radius
Number Of Known Moons	14
Planetary Ring System	6 Rings, 1 Containing Several Arcs